The Resilient Mindset

Resilience is a journey that shapes how we navigate challenges and grow through adversity. The *Resilient Mindset: Engineering Strength Through Systems Thinking* draws from the author's experiences in academic research, industry practice, and personal growth to explore resilience as a practical and transformative human process. By connecting the science of resilience with real-world challenges, this book offers a perspective on how individuals and systems alike can thrive under pressure and uncertainty.

This book delivers tools and insights designed to help individuals thrive in an ever-changing world. Readers will discover academic insights, practical tools, and lived experiences that illuminate resilience as both a mindset and a skill. Each chapter provides actionable strategies to enhance mental clarity, emotional adaptability, and long-term growth, supported by real-world examples and reflective exercises. It features examples of recovering from setbacks, adapting to changes, and sustaining progress in the face of uncertainty, through which it offers a clear and empowering path to building confidence and strength. Readers will learn how to navigate changes to the system and how they can stay resolute as processes and personnel change.

This book is for anyone seeking to grow through life's challenges, including human factors, professionals, health and safety personnel, human resource employees, and business and management practitioners.

The Resilient Mindset
Engineering Strength Through Systems Thinking

Mohammad Yazdi

CRC Press
Taylor & Francis Group
Boca Raton London New York

CRC Press is an imprint of the
Taylor & Francis Group, an **informa** business

First edition published 2026
by CRC Press
2385 NW Executive Center Drive, Suite 320, Boca Raton FL 33431

and by CRC Press
4 Park Square, Milton Park, Abingdon, Oxon, OX14 4RN

CRC Press is an imprint of Taylor & Francis Group, LLC

ISBN: 978-1-041-33803-1 (hbk)
ISBN: 978-1-041-33852-9 (pbk)
ISBN: 978-1-003-78989-5 (ebk)

DOI: 10.1201/9781003789895

Typeset in Times
by Apex CoVantage, LLC

For all those who stand for freedom and endure hardship,
for voices that refuse to be silenced,
for courage that transcends adversity,
and for the resilience that drives humanity forward.
This book is a tribute to the enduring strength of the human spirit.

Contents

Author Biography

Mohammad Yazdi's illustrious academic and professional journey began with his BSc in Safety and Technical Protection Engineering from the Petroleum University of Technology in Abadan, Iran, conferred in 2012. He further elevated his educational prowess by obtaining an MSc degree in Industrial Engineering from the Eastern Mediterranean University in Famagusta, Cyprus, in 2017. Immediately after, Mohammad integrated his expertise into academia by affiliating with the Centre for Marine Technology and Ocean Engineering (CENTEC) at the University of Lisbon in Portugal. He made significant contributions both in-person during 2017 and 2018 and remotely in 2019. His relentless pursuit of knowledge led him to undertake a dual Ph.D. program in 2022, partnering with the Centre for Risk, Integrity, and Safety Engineering (C-RISE) at the Memorial University of Newfoundland, Canada, and Macquarie University in Australia. Before venturing into these academic achievements, Mohammad dedicated himself to practical applications in the industrial world from 2012 to 2016. He took on multifaceted roles such as a Firefighter, Safety Officer, WHS/OHS Advisor, and Auditor, serving vital sectors including power plants, and the oil and gas industry. His research interests and professional background converge at the nexus of system safety, risk assessment, resilience, process integrity, and asset management, especially concerning renewable and non-renewable energy infrastructure. With an impressive track record of leading large-scale energy projects and technology-rich initiatives, Mohammad offers invaluable support to asset operators, developers, and maintainers.

Passionately, he endeavors to bridge the gap between industry and academia, crafting innovative solutions that foster advanced decision-making, systems thinking, and creativity. Mohammad champions an asset lifecycle approach from capital investment planning, spanning through operation, maintenance, and culminating in disposal. This holistic perspective allows him to integrate pragmatic thinking, balancing superior business outcomes with optimum functionality and reliability.

Preface

The wound is the place where the light enters you.

Rumi

Resilience is often marketed as a personal virtue or a motivational slogan. This book treats it as something more demanding and more hopeful: a way of understanding the distance between what life is and what we expect it to be, and a disciplined practice of building systems, technical and human, that can hold their shape when that distance suddenly widens.

In a world of unstable work, fragile institutions, and shifting routines, the question is no longer whether disruption will come, but what kind of inner and outer architecture we carry when it does.

The pages that follow grow out of three intertwined strands: formal research in safety and risk, years of practice in high-hazard industries, and lived experience of workplaces and systems that did not always behave as promised. Together, they suggest that resilience is not magic and not personality; it is design. It can be described, modelled, and, crucially, cultivated, in offshore platforms and power plants, but also in careers, relationships, and inner lives.

The aim of this book is to translate the language of resilience engineering and systems thinking into tools that can be carried into ordinary days: into difficult meetings, sleepless nights, layoff emails, complex projects, and quiet moments of doubt.

This is not a textbook in the traditional sense, and not a self-help manual. It is a bridge between two worlds: between the quantitative logic of performance curves and indices, and the qualitative texture of fear, hope, fatigue, and meaning. It treats human happiness as a dynamic relationship between "what is" and "what was expected", and asks how that relationship can be made less fragile, less dependent on everything going exactly to plan.

The Introduction, "The Resilient Mindset", lays the conceptual foundation. It introduces the central tension of the book – the gap between reality and expectation – and shows how much of our suffering grows in that gap when careers, organisations, or relationships fail to follow the script we had in mind. Resilience is framed here as a way of staying in contact with reality without surrendering dignity or direction. Drawing on examples from engineering and personal experience, the Introduction outlines resilience as a cycle of absorption, adaptation, recovery, and learning, and invites the reader to see both systems and selves through this lens.

Chapter 1, "Building the Bridge: Exploring Resilience Between Failure and Growth", develops a simple but powerful technical metaphor: performance functionality plotted against time. It shows how disruption, vulnerability, adaptation, and learning can be seen along a single performance–time curve, and proposes a straightforward Resilience Index for comparing how systems behave under shock without recourse to heavy mathematics.

Mechanical images – a tyre meeting a nail, a structure absorbing load – are placed alongside human scenes such as public speaking anxiety or career setbacks, so that readers can recognise the same pattern across steel and skin. The chapter closes by identifying four main levers for increasing resilience: reducing the size and frequency of shocks, shortening the time spent at low performance, supporting a stable recovery, and turning every disruption into a learning opportunity.

Chapter 2, "The Architecture of Flow: Minimising Disruptions to Performance", shifts the focus from major crises to everyday "micro-shocks". It treats human performance as a system that can be stabilised by managing common disruption domains: emotional, cognitive, physical, digital, social, and professional.

Drawing on neuroscience, psychology and organisational behaviour, the chapter shows how sleep loss, digital overload, workplace conflict, or shifting priorities slowly erode attention and reliability. For each domain, it distinguishes between proactive controls (habits, routines, boundaries) and reactive strategies (grounding, reframing, rapid triage), supported by short, practical examples from ordinary working days. The reader is guided through a simple exercise to identify their own most frequent disruption and design a preventive and a stabilising response.

Chapter 3, "The First Breath After Impact: Stabilising Before Recovery", moves into the immediate aftermath of a shock, the absorption phase. Here, resilience is described as the critical capacity to hold together when performance has dropped, emotions have surged and the nervous system is flooded with stress responses such as an amygdala hijack and cognitive overload.

Concepts from neuroscience, resilience engineering and job-demands/resources theory are used to explain why buffers, redundancy and simple grounding rituals matter so much in the first hours and days after impact. Across the same six domains introduced in Chapter 2, the book offers concrete stabilisation strategies: naming emotions, controlled breathing, single-tasking, gentle physical resets, digital boundaries, and social containment. Real-world stories of redundancy, relationship breakdown and workplace crises show how wise inaction, brief pauses, and containment can prevent long-term damage.

Chapter 4, "The Climb Back to the Ridge: Adaptation and Recovery in the Vulnerability Period", examines the long middle of resilience: the time after freefall has stopped but before stability has returned. This chapter distinguishes carefully between adaptation, the pivot that allows a person, organisation or infrastructure to function under new conditions, and recovery, the deliberate climb that rebuilds capacity towards an old or newly defined baseline.

Using examples from energy systems, pandemic response, and personal career disruption, it describes mechanisms such as sensemaking, prioritisation, guardrails, minimum viable routines, resource reallocation, and narrative reframing. The Resilience Activation Framework is introduced to show how different forms of capital (human, social, financial, institutional) can be mobilised during the vulnerability period, and how small, intentional steps shorten the window in which secondary shocks can do disproportionate harm.

Chapter 5, "Learning: Integrating Lessons for Future Resilience", closes and renews the cycle. It argues that resilience is incomplete without reflective learning, and that genuine "bouncing forward" requires more than survival: it requires insight.

The chapter explores reflective practices such as journaling, mindfulness, and after-action reviews, and explains how high-reliability organisations use blameless post-mortems and feedback loops to learn from both failures and everyday successes (Safety-II). At the societal level, it examines how communities can "build back better" after disasters by embedding learning into regulations, institutions, and shared narratives. By the end of the chapter, resilience has become an ongoing evolution: each cycle of disruption, absorption, adaptation, and recovery plants seeds for a higher level of capability, provided that lessons are consciously integrated rather than left implicit.

No book on resilience is written alone. The ideas here have been shaped by years of dialogue with students, colleagues, supervisors, friends, and workers in high-risk environments who trusted their stories to careful listening. In philosophical terms, they have been co-authors in practice: companions in inquiry who helped test concepts against lived reality, exposing where abstractions were too neat, and where models needed to bend to honour experience.

Gratitude is owed to the many classrooms, physical and virtual, where these ideas were first tried as sketches on a whiteboard or questions in a tutorial. Each question, challenge, and moment of shared silence acted as a small experiment in epistemology: what can really be known about resilience, and under what conditions? To colleagues and mentors in engineering, safety, psychology, and the humanities, thank you for widening the lens, for reminding us that every system has an ontology (what it is), a teleology (what it is for), and an ethics (who is affected when it fails).

Special thanks belong to those in industry and emergency services who live these questions under pressure: firefighters, safety practitioners, operators, technicians, and leaders who work in spaces where resilience is not a metaphor but a daily requirement. Their willingness to share near misses, hard nights, and quiet victories has given this book its practical backbone.

Finally, to family, friends, and the quieter circle of support who held space during times of disruption and doubt: your presence has been a form of living philosophy, an ethic of care that made it possible to treat adversity not only as a threat but also as a teacher. If this book succeeds in helping readers build lives and systems that are less brittle, more humane, and more attuned to the gap between expectation and reality, it will be because of the many minds and hearts that have walked alongside this work.

M. Yazdi
Sydney, December 2025

Foreword

This book is fundamentally about direction. Not speed. Not optimisation. Not heroic performance. Direction. It asks a deceptively simple question: *over time, which way is the system moving?* The answer, the book argues, determines whether work, organisations, and careers quietly strengthen, or slowly unravel. Its central arc is clear and deliberate:

> **Fragility → Stability → Recovery → Adaptation → Enduring Success**. This arc mirrors how real systems behave in the presence of stress, uncertainty, and imperfect decision-making.

The book opens by confronting **fragility**, the most underestimated condition in modern work. Fragility rarely announces itself. It hides behind apparent efficiency, impressive output, and short-term wins. Morgan Housel often notes that ruin usually arrives not through dramatic mistakes, but through a long series of reasonable decisions made without sufficient margin. The same dynamic plays out in workplaces:

> **Tight deadlines → skipped recovery → rising cognitive load → normalised shortcuts → eventual breakdown**.

A team meets targets by working longer hours; a manager defers maintenance to hit quarterly numbers; a professional relies on memory instead of documentation because "it's faster". Each choice feels rational in isolation, yet collectively they move the system towards a cliff. The book excels at making this invisible trajectory visible.

From fragility, the book deliberately redirects the reader towards **stabilisation**, emphasising the power of modest, conservative actions:

> **Efficiency obsession → buffers → slack → operational stability**.

This logic closely aligns with Warren Buffett's long-standing principle of *margin of safety*. Buffett's success was never about predicting the future better than others; it was about avoiding situations where a single error could cause irreparable damage. In practical terms, the book translates this idea into everyday work: cross-training instead of single experts, spare capacity instead of zero-tolerance scheduling, and redundancy instead of brittle optimisation. A single backup operator, an extra day in a project plan, or a simple escalation rule can dramatically alter outcomes, not by increasing brilliance, but by reducing vulnerability.

Once stability is established, the arc moves into **recovery**, which the book treats as a core performance metric rather than a post-failure inconvenience:

Disruption → rapid restoration → preserved confidence → learning.

Here the parallels to long-term investing become even clearer. Buffett avoids leverage not because downturns are impossible to predict, but because recovery becomes impossible once flexibility is lost. Similarly, the book shows that systems which recover quickly, after outages, errors, or overload, outperform systems that merely try to avoid failure. A software team that can roll back deployments in minutes, a hospital unit that restores workflow after a surge, or a professional who protects sleep and decision capacity during intense periods all share one advantage: they remain in the game.

The next phase of the arc is **adaptation and compounding advantage**:

Stability → reflection → adjustment → incremental gains.

This is where Morgan Housel's insights into compounding are particularly relevant. Just as financial returns compound quietly over decades, resilience compounds through small, repeated practices. One checklist prevents one error → one avoided error prevents rework → preserved energy improves judgement → better judgement improves the next decision. None of these steps are dramatic. Yet, over years, they separate systems that decay from those that strengthen. The book repeatedly reinforces this point with practical mechanisms: short post-task reviews, simple leading indicators of strain, and routines that turn experience into improvement rather than fatigue.

As the arc approaches its final stage, the book reframes **success** itself:

Short-term performance → sustained performance → long-term success.

Success is no longer defined by uninterrupted growth, constant urgency, or flawless execution. Instead, it is defined by endurance, by the ability to absorb shocks without collapse, recover without panic, and adapt without losing identity. This mirrors Buffett's observation that *the first rule of compounding is to avoid interrupting it.* In work and life, burnout, catastrophic failure, and loss of trust are the interruptions that matter most. The book shows how resilient design prevents those interruptions before they occur.

What makes this book especially powerful is that it does not demand radical transformation. It encourages readers to move along the arc intentionally, one decision at a time:

> **Remove one single point of failure → add one buffer → measure one recovery time → protect one recovery habit**.

Each small step changes the direction of travel. Over time, direction dominates intensity. Fragile systems may look impressive briefly, but resilient systems endure.

Ultimately, this book makes a quiet but persuasive argument: the principles that govern long-term financial success, margin of safety, avoidance of ruin, and compounding through consistency, also govern sustainable work, leadership, and life design. Resilience is not the opposite of ambition. It is the condition that allows ambition to endure.

For readers seeking durability rather than drama, progress rather than heroics, and success that survives uncertainty, this book provides not just insight, but a map. The arrows are clear. The direction is intentional. And the destination, enduring, resilient success, is both achievable and worth the discipline it requires.

Dr He Li
Liverpool, UK
January 2026

Introduction

Even the darkest night will end and the sun will rise.

Victor Hugo

Resilience is more than a motivational slogan. It is a way of seeing the world and a way of moving through it. It is not limited to "bouncing back" after hardship; it is the capacity to stay in contact with reality when reality does not match the script and still act in a way that protects what matters.

At the centre of this mindset sits a simple tension: the distance between what life is and what was expected of it. Much of human unhappiness grows in that gap. When the difference is small, the mind calls it an inconvenience. When the difference is large and long-lasting, the mind calls it failure, injustice, or loss. Resilience does not pretend that gap away. It changes how that distance is understood and how people respond to it.

This book brings together three strands: research in safety and risk, practice in high-hazard industries, and lived experience in imperfect workplaces and systems. Together, they point to the same conclusion: resilience can be engineered, not only in offshore platforms and industrial plants, but also in careers, relationships, and inner lives.

The aim is to translate systems thinking and resilience engineering into practical tools that help people:

- stay grounded when expectations are shattered,
- adjust plans without abandoning their values,
- and design their lives in ways that make happiness less fragile and less dependent on everything going exactly as planned.

WHERE IT ALL BEGAN

My journey into resilience started during my PhD in 2019, right in the midst of the coronavirus disease of 2019 (COVID-19) pandemic. The initial impulse for this work came from engineering problems, not self-help questions. Offshore

DOI: 10.1201/9781003789895-1

structures exposed to corrosion, waves, and uncertain loading conditions must continue to function in spite of degradation, delayed maintenance, and changing environments. Traditional approaches focused on preventing failure through stronger materials and more inspections. Towards the end of my thesis, encountered a concept that changed everything: *resilience engineering*. Resilience engineering added a different question: how can a system continue to perform acceptably even when many things are going differently than expected?

Asking deeper questions:

> How can we design offshore structures to absorb shocks, adapt to unexpected events, recover quickly, and extend their lifespan?

That shift in perspective, away from a fantasy of perfect control towards the design of systems that can live with imperfection, turns out to be just as relevant to human life as to steel and concrete. Life rarely follows the technical drawing. Careers that were supposed to be linear become jagged. Organisations that promised stability restructure overnight. Economies lurch, families fracture, and health fluctuates.

That insight sparked a bigger question:

> *If a structure can be resilient, can an individual be too?*

Resilience engineering introduced useful language: absorption, adaptation, recovery, and learning. These words describe how systems cope with disturbance. They also describe how people cope with disappointment.

The idea for this book grew from my engineering roots and research mindset. I fell in love with the concept of resilience because it's universal, it applies not just to structures, but to people, organisations, and societies. The goal is to share these insights in a way that resonates with a broader audience, beyond the engineering world.

FROM THEORY TO EXPERIENCE

Abstract models of resilience only become meaningful when they are tested against ordinary, painful situations: a contract that ends suddenly, a promotion that never comes, a country that no longer feels like home, or a relationship that quietly empties out long before it officially ends.

In such moments, a person encounters two things at once:

1 What actually happened and
2 What was expected to happen.

The distance between these two is often where suffering lives. A job loss hurts not only because work has stopped, but because a story about the future has been broken. A difficult workplace crushes morale not only because of the tasks, but because the environment clashes with what was believed about fairness, respect, or recognition.

Over time, however, patterns emerge. Some people remain trapped in the first shock, repeating the story of what "should have" happened. Others slowly develop a different stance. They still feel the loss, but they start asking different questions:

- What does this reveal about the systems involved, not just about personal worth?
- Which expectations were realistic, and which were quietly imported from someone else's script?
- What can be redesigned: skills, boundaries, networks, and routines?

Resilience, in this sense, is not built in a single dramatic breakthrough. It is assembled across many encounters with unwanted reality. The more honestly those encounters are examined, the more structured and less random that assembly becomes.

From my experience, after completing my PhD, publishing many journal papers, book chapters, and books in the field of critical system safety, I felt confident and ready to take on the world. My first job was at a consultancy firm in Sydney. Just two months in, I was let go due to poor performance. It was shocking and I cried a lot. I started to question everything. It took weeks to process what had happened. Even now, thinking back to those days brings a wave of sadness.

Eventually, I had to move on and joined the events industry. It was a fun, vibrant, fast-paced environment full of energy. But after a year, I was made redundant. The human resources (HR) meeting when I heard the news still stays in my memory. I felt fine at first and stayed calm, but the reality sank in later. Again, time moved on. Then came another job, which I left after just two weeks due to a toxic work culture. This time, I walked away with confidence, clarity, pride, and no fear.

Looking back, I began to ask myself some important questions. *What shifted inside me? How did I go from feeling completely broken after my first job loss to calmly walking away from a toxic workplace with confidence?* Something had clearly changed. I was no longer reacting with fear or self-doubt. Instead, I was making decisions with clarity and strength. I started to realise that I had developed a new kind of inner capacity. I had become more adaptable, more grounded, and more focused on moving forward rather than staying stuck in disappointment. This was resilience in action, lived, not theorised.

This transformation didn't happen overnight. It was shaped by experience, reflection, and a willingness to grow through discomfort. Each setback became a lesson. Each challenge became a test of how I could respond, not just emotionally, but strategically. I began to see patterns in my behaviour and was learning how to absorb the impact of unexpected events, how to adjust my mindset, and how to return to my goals with renewed energy. This was resilience in action.

I didn't read it in a book. I didn't attend a workshop that gave me a magic formula. I absolutely lived it. And through that process, I started to understand that resilience is something we build through practice. It is a skill, a mindset, and a way of approaching life that allows us to keep moving, even when the path is uncertain.

EXPECTATIONS, HAPPINESS, AND THE GAP BETWEEN

Happiness is often treated as a mysterious emotion, but in everyday life it behaves more like a relationship: a relationship between what is possessed and what is expected.

- When reality rises above expectation, the result feels like joy or gratitude.
- When reality roughly matches expectation, the result feels like calm or contentment.
- When reality falls far below expectation, the result often feels like anger, shame, or despair.

The same objective situation can feel very different depending on the internal reference point. A modest apartment can feel like abundance to someone who expected nothing and like humiliation to someone who expected a waterfront house. The external situation did not change; the expectation did.

Modern culture quietly trains very high, very fragile expectations: permanent career growth, constant productivity, uninterrupted emotional comfort, and a life that reads like a carefully edited highlight reel. When those expectations are taken as the default, almost any real life will feel like a disappointment.

Resilience does not advocate low standards or passive acceptance. Instead, it encourages clear, robust expectations:

- Expect that plans will change, and design with slack.
- Expect that other people are inconsistent, and build in boundaries rather than fantasies.
- Expect that some suffering is part of any serious commitment, whether in work, learning, or love.

When expectations acknowledge how systems behave, happiness no longer depends on perfection. Small gains are not invisible next to impossible standards; they are recognised and valued. Setbacks still hurt, but they are interpreted as part of a longer process rather than a final verdict.

A resilient mindset continually adjusts this internal reference point. It asks, "Given the world as it is not as it was promised what is a good, meaningful next step?" That question is less fascinating than grand visions of success, but it is far more stable. It narrows the destructive gap between fantasy and reality without erasing ambition or hope.

A MESSAGE THAT RESONATED: LEARNING FROM "PAIN AND SUFFERING"

Jensen Huang, the Chief Executive Officer (CEO) of NVIDIA, has spoken openly about "pain and suffering" as ingredients of durable success. His message is straightforward: comfort does not teach us how to suffer, and without learning how to suffer, people do not learn how to be resilient. He told students: *You're lucky to be here. You'll probably be successful because of your education and your network. But most of you are here because you grew up in comfort. And comfort doesn't teach you how to suffer. Without suffering, you don't build resilience.*

The early history of NVIDIA is a sequence of near-failures, dead ends, and hard resets. Those episodes did not simply test the company; they shaped its identity. They forced clarity about what mattered, discipline in execution, and patience in the face of delayed rewards. Huang has consistently emphasised that resilience is the foundation of both personal growth and organisational strength. He believes that resilience is shaped through hardship, adversity, and the ability to endure. Intelligence and privilege may open doors, but they do not guarantee the strength to walk through them when the path becomes difficult.

The fact is that people with very high expectations can struggle when reality doesn't meet those expectations. Character is what carries us forward, and character is built through challenge. Within NVIDIA, Jensen frequently uses the phrase "pain and suffering" to describe the process of growth. He recalls how the company faced multiple existential threats in its early years, and how overcoming those crises helped shape its culture and long-term resilience.

Here, again, happiness links back to expectations. If the expectation is an easy, linear, painless life, almost everything will feel like an insult. If the expectation is that meaningful work and relationships will involve effort, conflict, waiting, and loss, then moments of ease are received as gifts, not entitlements, and difficulty is less likely to be interpreted as proof that something is fundamentally wrong.

This philosophy helped me reframe my own setbacks. I began to see that the difficult moments in my life were not obstacles to avoid, but indeed experiences that helped me grow stronger. They taught me how to recover, how to adapt, and how to move forward with greater clarity. Resilience became something that could be felt within myself, a quiet strength that allowed me to keep going, even when things felt uncertain.

Huang also extends this idea beyond individuals. He speaks about resilience in global supply chains, especially in the face of geopolitical tensions. His message is clear: resilience is essential for long-term success, whether in personal life, company culture, or international operations. It lives in the mindset of people who are willing to face problems with determination and ask themselves, *How hard can this be?*

This perspective helps us to understand that resilience is something we can build. It grows through experience, reflection, and the choices we make when things don't go as planned. Each challenge becomes an opportunity to strengthen our foundation. Each setback becomes a lesson in endurance. And each decision to keep moving forward becomes a step towards becoming more capable, more confident, and more resilient.

THE HUMAN SYSTEM

Human beings can be viewed as complex systems with inputs, outputs, and feedback loops. This view does not reduce a person to machinery; it simply gives language for patterns already felt:

- Some lives run on chronic overload: too much input, no time to process, no margins.
- Others run on chronic under-load: too little challenge, unused capacity, slow erosion of motivation.
- Many are governed by feedback loops in which each disappointment reinforces a story of inadequacy, increasing the likelihood of the next disappointment.

Within this system, happiness and resilience depend on several capacities:

- *Buffering capacity* – enough time, energy, relationships, and savings to absorb shocks.
- *Adaptive capacity* – the ability to change strategies when information or conditions change.
- *Recovery capacity* – routines that restore attention, physical health, and emotional balance.
- *Learning capacity* – the habit of extracting structure from experience, not just surviving it.

Expectations sit inside this system as powerful internal parameters. Unrealistic expectations tighten feedback loops of frustration. Clear, flexible expectations loosen those loops, making it easier for new information to change behaviour instead of just confirming old fears.

Seeing oneself as a system does not remove mystery from life, but it does reduce confusion. Instead of asking only, "Why am I like this?", the question can shift to, "What kind of system am I running, and which parts of it can be redesigned?"

Engineered systems are designed to withstand shocks, recover from disruptions, and perform under pressure. Similarly, we can develop these same qualities within ourselves. Through practice, awareness, and deliberate effort, we can train our minds to stay calm, strengthen our emotional responses, and build habits that support recovery and growth.

This process begins with awareness. When we understand how we function as systems, we can begin to identify the areas that need reinforcement. We can examine how we react to setbacks, how we process feedback, and how we make decisions under pressure. From there, we can introduce tools, strategies, and routines that help us become more adaptable and more capable of bouncing back.

Resilience is more than recovery. It includes preparation, anticipation, and transformation. It allows us to face challenges with confidence, to learn from adversity, and to emerge stronger and more focused. It gives us the ability to maintain our core purpose even when circumstances shift, and to evolve when growth is needed.

A ROADMAP FOR THIS BOOK

This book explores resilience as both a mindset and a practice. Each chapter breaks down the concept into practical steps, offering tools to strengthen mental clarity, emotional agility, and long-term growth. You will find real-world examples, strategies that can be applied immediately, and reflections that encourage deeper self-awareness.

Everything presented here comes from recent academic research, published works, and years of experience in both professional and personal settings. The purpose is to share insights that are practical, evidence-based, and directly useful.

Whether you are facing personal challenges, moving through career transitions, or simply seeking to become more grounded and adaptable, this book is designed to guide you step by step towards building resilience with clarity and intention.

Building the Bridge

1

Exploring Resilience Between Failure and Growth

Let everything happen to you: beauty and terror. Just keep going. No feeling is final.

Rainer Maria Rilke

This chapter touches only lightly on engineering, just enough to reveal the quiet forces that hold resilience together. Every system carries an interior logic, a hidden rhythm that shapes its response to stress. Once that rhythm is understood, the broader narrative can unfold without the weight of technical detail, guided instead by clarity and insight.

One of the challenges today is that we often lack clarity about the foundational concepts: what resilience *is* and where the "battlefield" lies. The battlefield isn't always obvious. We often find ourselves juggling metaphors, switching between terminologies, and navigating frameworks that don't quite align. It's like trying to build a bridge between the ground and the sky, between gritty, real-world disruptions and lofty, abstract models. And sometimes, that bridge feels shaky.

So let's ask something.

Imagine you're driving through the vast, sunburnt desert near Alice Springs. The horizon stretches endlessly, the air shimmers with heat, and the road ahead is quiet and unforgiving. The engine hums, the radio murmurs in the background, and for a moment everything feels steady and under control.

DOI: 10.1201/9781003789895-2

Then, without warning, there's a dull thud, a pull on the steering wheel, and that heavy sinking feeling in your stomach. The car starts to drag. You know, even before you stop, that it's a flat tyre. You roll onto the narrow shoulder and step out into the heat. It hits you like a wall. There's no shade, no buildings, no nearby town, just open land and a strip of bitumen cutting through it. The silence is almost too loud. No other cars in sight. You check your phone; the signal is weak or gone. For a moment, you feel very small. What seemed like a simple, manageable trip now feels fragile. Your mind jumps quickly: Do I have enough water? How long until someone drives past? What if the spare is no good? What if no one knows I'm out here? The day you had planned, the meeting you were heading to, the people expecting you, the quiet confidence that you would just get there, starts to fall apart in your head. Just like that, your journey grinds to a halt. The car is still there, the road is still there, but the sense of certainty is gone. A small mechanical failure has turned into a chain of worries. You're stranded. Plans unravel. The simple goal of getting from point A to point B has just failed, completely and unequivocally, and you are left standing in the heat, feeling the full weight of that failure.

Now picture a different version of that moment.

Same desert. Same road. Same nail on the ground. But this time, your tyre is equipped with an anti-puncture sealant. The moment the nail pierces the rubber, the sealant kicks in quietly, and automatically. The tyre doesn't go flat. You keep driving. Maybe a little slower, maybe with a slight wobble, but you're moving. The journey continues. The system absorbed the shock, adapted, and recovered without collapsing.

That's resilience.

What's the difference between these two situations? The first is a failure. The second is resilience. It's not about perfection. It's about persistence. It's about systems that bend without breaking, that absorb disruption and keep going. And understanding *how* that happens – mechanically, conceptually, and metaphorically – is the key to everything that follows.

Now, shift the lens from mechanical systems to something more personal: human growth.

Imagine someone stepping into the daunting world of public speaking. Their first few attempts are rough. Their voice trembles, they forget their lines, and the feedback they receive stings. The experience is uncomfortable, even humiliating. And if they walk away at this point, if they decide it's not for them, that's a kind of failure – not because they weren't talented, but because their internal system, their confidence, their preparation, their emotional resilience, couldn't absorb the stress. It cracked under pressure.

Now imagine someone else in the same situation. They also stumble. Their hands shake, their words falter. But instead of retreating, they pause. They reflect. They ask, *What went wrong? What can I learn?* They seek feedback, not as judgement, but as fuel. They practice deliberately, even when it's uncomfortable. They still feel nervous, but each time they speak, something shifts. A little more control. A little more clarity. A little more courage.

Their system – mental, emotional, and psychological – is adapting. It's learning to absorb the discomfort, to metabolise the stress, and to grow stronger because of it.

That's resilience.

Resilience begins where struggle lives. It's the choice to remain engaged when things get hard, to keep showing up even when progress feels slow, and the path ahead looks steep. It's the quiet courage to continue, to lean into discomfort and grow through it. Resilience embraces challenge, meets it head-on, and transforms adversity into strength. It's the art of turning discomfort into transformation, step by step, moment by moment.

Resilience, in the context of system performance, is a journey shaped by disruption, response, and renewal. A system begins with reliable functionality, operating smoothly and fulfilling its intended purpose, until it fails. Why? Because the reliability of any system will be decreased over time. But when a disruption occurs, that reliability is challenged. The system faces a turning point: either it continues to function in some capacity, or it descends into failure.

This moment of disruption exposes the system's *vulnerability*. The degree to which performance drops reflects how well the system can *absorb* the shock. Absorption is the first layer of resilience; it determines whether the system can resist immediate collapse and maintain some level of operation.

Following this, the system enters a phase of *recoverability*, shaped by its ability to *adapt* and *restore* functionality. Adaptation involves adjusting internal processes, reconfiguring resources, and finding new ways to operate under changed conditions. Restoration is the gradual return to a stable state, in which performance begins to climb again.

As the system stabilises, *learning* becomes essential. This is where resilience deepens. The system reflects on the disruption, integrates lessons, and evolves. Learning enhances future responses, making the system more capable of facing similar challenges ahead.

These elements unfold across two essential dimensions: *Time* and *Performance Functionality (PF)*.

- *Time* represents the duration over which the system experiences disruption, responds to it, and regains stability. This can be measured in any unit: seconds, hours, or days, depending on the context.
- *PF* reflects the system's operational capacity before, during, and after the shock. It can be expressed as a percentage, where full functionality equals 100%, and lower values indicate reduced performance.

Together, these dimensions form the canvas on which resilience is painted. A system begins with reliable performance. Then, a disruption occurs, an unexpected event that challenges its stability. If the system cannot absorb the shock, performance drops, and failure becomes a real possibility. What happens next depends on the system's ability to adapt, recover, and learn.

We've provided an illustration to show this process (Figure 1.1), along with a simplified mathematical formulation to support clarity. The *Resilience Index (RI)* offers a way to quantify how well a system manages change across these dimensions. We present the formulation here for reference, without further discussion:

$$\overline{RI} = PF(Reliability \mid Disruption) + PF\left(Absorption\right) + PF\left(Adaptation\right) + PF\left(Restoration\right) + \left\{PF\left(Learning\right)\right\}$$

Resilience lives in the space between disruption and recovery [1–2]. Whether it's a tyre absorbing the shock of a nail on a desert road, or a person trembling through their first public speech, the essence remains the same: something is challenged, and something responds. The system, mechanical or human, doesn't collapse. It absorbs, adapts, and begins to recover [3]. Over time, PF shifts. It may dip, fluctuate, or slowly rise, but the journey continues. Time marks the rhythm of this transformation, and performance reflects its depth. We've provided a figure and a formulation to support this understanding,

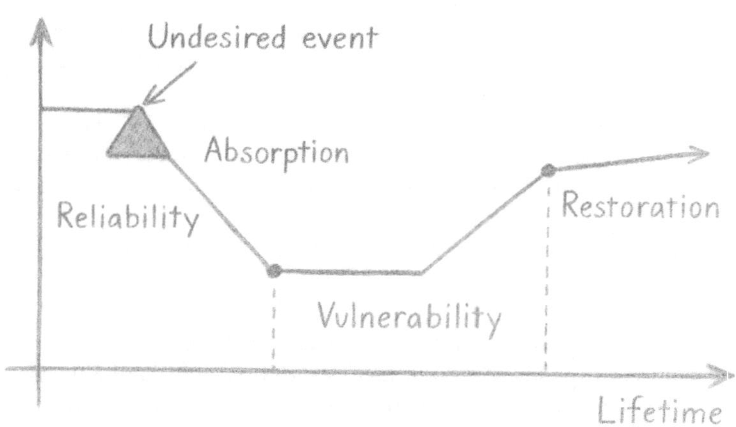

FIGURE 1.1 Performance functionality over the system's lifetime.

but the heart of resilience is simpler: it sustains purpose through uncertainty, adjusts capacity in response to stress, and evolves through experience. It's the quiet strength that keeps a system moving forward, even when the path is disrupted [4].

Now that we've explored how resilience works, step by step, across time and performance, we can begin to see the full picture. Whether it's a tyre absorbing a puncture in the desert or a person trembling through their first public speech, resilience is the quiet force that keeps systems, and people, moving forward. It's the success of the individual over time, shaped by the system's inherent capacity to endure, adapt, and grow.

We've already discussed that resilience is an embedded feature of the system itself. Just think back to Jensen Huang's reflections on suffering and pain, how challenge becomes the crucible for strength. The question now is: how do we improve resilience? How do we become more resilient?

As you've seen, resilience unfolds across two key dimensions: *Time* and *PF* [5]. Within these dimensions, there are four practical strategies for strengthening resilience:

1 *Reduce the number of shocks*: Every disruption carries a cost. The fewer shocks we encounter, the more smoothly we can operate, and the more energy we preserve for growth. Think of a person navigating a demanding work environment. If they're constantly bombarded by unexpected crises, tight deadlines, interpersonal conflict, or shifting expectations, their ability to stay focused and perform well begins to erode. But when those shocks are reduced, through better planning, clearer communication, or supportive leadership, their performance stabilises. They can breathe, think, and thrive. Reducing shocks doesn't mean avoiding challenge, it means creating space for resilience to flourish.

2 *Minimise the vulnerability period*: When disruption does occur, the speed at which we absorb it matters. A resilient person doesn't avoid stress; they learn to metabolise it quickly. Imagine someone receiving difficult feedback after a presentation. One person might spiral into self-doubt for days, while another takes a moment to feel the sting, then begins to reflect, recalibrate, and move forward. The second person shortens their vulnerability period. They preserve their core strength by responding with clarity and intention. The faster we absorb the impact, the less damage it does, and the more control we retain.

3 *Shorten the recovery phase*: Recovery is where resilience becomes visible. It's the climb back from disruption, shaped by adaptation and restoration. Picture an athlete who suffers an injury. Their

recovery depends on how quickly they adapt, modifying their training, adjusting their mindset, and how effectively they restore their physical capacity. The shorter the recovery phase, the sooner they return to performance. In human terms, this means building habits that support healing, cultivating flexibility, and staying connected to purpose even when progress feels slow.

4 *Strengthen learning*: Learning is the heartbeat of resilience. It's what transforms a setback into a stepping stone. When someone reflects on a failure, asks *what went wrong*, and integrates that insight into future action, they're not just recovering, they're evolving. Learning influences every part of the system. It improves absorption, accelerates adaptation, and enhances recovery. It's the mechanism through which resilience becomes sustainable. Without learning, we repeat patterns. With it, we grow stronger, wiser, and more prepared.

Resilience is never just about bouncing back. It's about moving forward with greater depth, greater clarity, and greater strength. These strategies – reducing shocks, minimising vulnerability, shortening recovery, and deepening learning – are the tools we use to shape that journey. And in the end, resilience is the story of becoming, of turning disruption into transformation, and challenge into character.

In the next chapters, we'll take a closer look at human system resilience, one layer at a time. We'll explore the most applicable approaches drawn from best practices across disciplines, with the goal of helping each of us take at least one meaningful step forward. This isn't just theory; it's an absolute journey. And if you're serious about building resilience, it requires commitment. As the saying goes, *if you want to take the island, you must be willing to burn the boat.* There's no turning back. Resilience grows when we fully engage with the process, not as a trial run, but as a lived experience.

We'll move from understanding to action. You'll learn how to strengthen your own system mentally, emotionally, and behaviourally by applying strategies that reduce shocks, shorten vulnerability, accelerate recovery, and deepen learning. These are not abstract ideas; they're practical tools for real life.

To support this, each chapter will end with a *Quick Practice*, a simple, focused exercise designed to help you apply what you've learned. These practices won't require hours of effort or perfect conditions. They're meant to be accessible, repeatable, and powerful in their simplicity, because resilience grows through small, consistent steps.

So, as we move forward, remember this is your journey, and every step you take strengthens the system within you.

QUICK PRACTICE: MAPPING YOUR RESILIENCE SYSTEM

Step 1: Identify a Recent Disruption

Think of a moment in the past week or month when something didn't go as planned. Write it in a single sentence.

My Disruption: ________________________________
(Example: I froze during a team presentation and forgot what I was supposed to say.)

Step 2: Reflect on Your Response

Answer briefly – be kind to yourself; this is reflection, not judgement.

- *How did I absorb the impact?*
- *How long did I feel vulnerable?*
- *What helped me begin to recover?*
- *What did I learn from the experience?*

Step 3: Map Your Resilience Dimensions

On a timeline, mark key phases of the disruption. Then estimate your *PF* in percentages:

- When the disruption occurred
- How long I felt vulnerable
- When recovery began
- When stability returned

Step 4: Choose One Adjustment

Based on your reflection, choose one small change for next time.

My Commitment: ________________________________
(Example: Next time, I'll rehearse my opening lines three times the night before.)

Tip: Repeat this exercise with different disruptions over time. Small adjustments build strong and sustainable resilience.

This practice is your starting point. It's not about perfection, it's about progress. Each time you reflect, adjust, and act, you strengthen your system. Resilience grows through repetition, awareness, and intention.

REFERENCES

1. Herrman, H., Stewart, D.E., Diaz-Granados, N., Berger, E.L., Jackson, B. and Yuen, T., 2011. What is resilience? *The Canadian Journal of Psychiatry*, 56(5), pp.258–265.
2. Wu, G., Feder, A., Cohen, H., Kim, J.J., Calderon, S., Charney, D.S. and Mathé, A.A., 2013. Understanding resilience. *Frontiers in Behavioral Neuroscience*, 7, p.10.
3. Manyena, S.B., 2006. The concept of resilience revisited. *Disasters*, 30(4), pp.434–450.
4. Masten, A.S. and Reed, M.G.J., 2002. Resilience in development. *Handbook of Positive Psychology*, 74(2), pp.74–88.
5. Martin-Breen, P. and Anderies, J.M., 2011. *Resilience: A literature review*, Bellagio Initiative, IDS.

The Architecture of Flow 2

Minimising Disruptions to Performance

The only journey is the one within.

Rainer Maria Rilke

Imagine your day unfolding: the alarm rings after a fitful night; you glance at a flood of urgent emails and feel your heart rate spike; mid-morning, you're interrupted by a message, and by afternoon, an argument at home remains on your mind. Each small shock – emotional tension, a barrage of digital notifications, and a lapse in sleep – chips away at your ability to perform at your best. Resilience often recalls "bouncing back", but an equally vital strategy is minimising disruptions in the first place. In other words, we want to build our personal system so that it (like an engineered system) attracts fewer shocks and stays on course. This means anticipating common stressors and putting safeguards in place, as well as having tools ready to respond when surprises do come.

Our brains and bodies simply cannot handle constant turbulence. Neuroscience shows that even brief interruptions dramatically slow us down: the brain isn't optimised for multitasking but functions best during deep, uninterrupted focus [1]. Every ping, every sudden demand, forces a costly "reorientation" as the mind switches tasks. In fact, research finds that a quick glance at a social media update or an email notification can significantly lengthen task

DOI: 10.1201/9781003789895-3

completion times [2]. Over time, these micro-disruptions accumulate, eroding concentration and draining energy. Meanwhile, other forms of disruption – emotional stress, poor sleep, and social conflict – trigger hormonal and cognitive changes that blunt our thinking. For example, lack of sleep has been shown to impair memory, attention, alertness, and decision-making, leading to "decreased function and impaired cognitive performance" [3]. Elevated stress hormones like cortisol similarly undermine mental flexibility and working memory.

To perform reliably, we must thus attack disruptions on all fronts: emotional, mental, physical, digital, social, and professional (Figure 2.1). Think of human performance like a symphony: each domain is an instrument, and when one falls out of tune, the entire composition suffers. In the following sections, we highlight the most critical categories of disruption, those that most commonly derail human functionality and performance. These are the "undesired events" which act like sudden gusts of wind against a carefully charted sailboat, threatening to push us off course:

- Emotional Well-being (Emotional Disruptions)
- Mental Focus (Mental/Cognitive Disruptions)
- Physical Well-being (Physical Disruptions)
- Digital Distractions (Technology Disruptions)
- Social Environment (Social Disruptions)
- Professional Context (Workplace Disruptions)

Each of these domains represents a potential fault line – places where stress, distraction, or conflict can crack the foundation of our performance. To treat them effectively, we must first understand their nature and impact. Then, we must identify the best course of action for each case. As the saying goes, *An ounce of prevention is worth a pound of cure.* This is especially true in performance science, where early intervention can prevent cascading failures.

Throughout this chapter, we draw upon findings from outstanding scientific research, studies that have conducted practical and feasible investigations into how these disruptions manifest and how they can be mitigated. These insights stem directly from neuroscience, psychology, organisational behaviour, and human factors engineering. Like a pilot reaching for a flight manual during turbulence, we rely on these studies to navigate real-world shocks with precision and confidence.

EMOTIONAL WELL-BEING (EMOTIONAL DISRUPTIONS)

Emotions have a way of reshaping the entire mental landscape. They don't just whisper in the background, they arrive like sudden storms, scattering focus and clarity in their wake. Anxiety, fear, anger, or sadness aren't passive states; they actively influence how thoughts form, how decisions are made, and how performance unfolds. Consider the moment when unexpectedly harsh feedback lands on work that consumed weeks of effort. The words themselves might be few, but their weight lingers, echoing long after the conversation ends. Suddenly, the day feels heavier. Motivation dips, thoughts spiral, and even the simplest tasks seem to demand more energy than they should. It's a reminder that emotions aren't just feelings, they're forces that shape behaviour, productivity, and perspective in ways that are hard to ignore.

Table 2.1 provides a summary of the main disruption domains, their typical triggers, observable indicators, and practical control options.

These emotional shocks trigger a surge of physiological responses. Stress hormones surge, and the amygdala, the brain's emotional alarm system, takes control, sidelining the prefrontal cortex which governs logic, planning, and decision-making. In this state, concentration falters, memory blurs, and composure slips.

As Maya Angelou once said, *You may not control all the events that happen to you, but you can decide not to be reduced by them.* Emotional resilience begins with recognising the storm and choosing how to steer through it. Understanding the science behind emotional disruptions and learning to respond with intention, we reclaim our ability to perform, even under pressure.

- *Proactive Strategies:* Emotional resilience grows through daily rituals, like watering a plant before the heat of the day. I have found that the most effective way to stay grounded is to build habits that strengthen my emotional core. Mindfulness and emotional awareness have become essential tools. Research in neuroscience and psychology shows that present-moment, non-judgemental awareness helps reduce stress and sharpen attention [4]. Each morning, we spend a few minutes in mindful breathing. This simple act helps us catch anxiety early, before it takes root, and allows us to move through the day with greater clarity.

TABLE 2.1 Everyday disruption domains, observable indicators, simple metrics, and control actions

DISRUPTION DOMAINS	*EXAMPLE WORKPLACE TRIGGERS*	*OBSERVABLE INDICATORS IN PERFORMANCE*	*SIMPLE METRICS (EXAMPLE DATA)*	*PREVENTIVE CONTROLS (BEFORE DISRUPTION)*	*STABILISING ACTIONS (AFTER DISRUPTION)*
Emotional	Harsh feedback, conflict with supervisor or colleague, job insecurity, constant criticism	Rumination, loss of focus, avoidance of tasks, mood swings, increased error rate on routine work	Self-reported stress (0–10) at end of shift (e.g., baseline 3–4, disrupted week 7–8); minor error reports per week (baseline 1–2, spike to 6)	Regular check-ins with leader, psychological safety norms, clear expectations, access to employee assistance programmes/peer support, training in basic emotional regulation	Short reset break (5–10 min walk), naming the emotion, brief debrief with trusted colleague, re-prioritising workload, deferring non-critical tasks
Cognitive/ mental load	Multitasking, frequent task switching, urgent requests, complex decision-making under time pressure	Slower decisions, difficulty prioritising, forgetting small tasks, rework, difficulty following procedures	Task switches per hour during focus work (e.g., baseline 4–6, disrupted 15+); rework rate (%) on key tasks (baseline 3–5%, disrupted 12%)	Protected focus blocks, work-in-progress limits, clear escalation rules, visual task boards, simple decision aids/ checklists	Pause– prioritise –plan (3–minute triage), breaking work into smaller steps, asking for temporary help, using a written checklist to offload memory

Physical/ fatigue	Long shifts, poor sleep, physical strain, heat or cold, repetitive manual handling	Yawning, slower reaction times, micro-errors, near misses, reduced participation in discussions	Average sleep hours/ night (baseline 7.0–7.5 h, disrupted week 5.0–5.5 h); near misses per month (baseline 0–1, spike to 4–5)	Fatigue risk management (shift design, breaks), ergonomic setup, manual handling aids, hydration/thermal comfort controls	Short micro-breaks (2–3 min), stretching, task rotation, reassignment from high-risk tasks, early finish/recovery period if risk is high
Digital/ technology	Constant emails, chat notifications, multiple systems, poorly designed interfaces, alarms	Fragmented attention, delays due to "chasing" information, wrong data entry, incomplete records	Interruptions per hour from digital tools during core focus time (baseline <4, disrupted 10–15); time spent searching for information per task (baseline 1–2 min, disrupted 6–8 min)	Notification rules, agreed "no-email" focus windows, interface redesign, standard naming and filing, training on core systems	Muting non-critical alerts for a fixed period, using "single screen" mode, batching communication tasks, quick call to clarify instead of long email chains

(Continued)

TABLE 2.1 Everyday disruption domains, observable indicators, simple metrics, and control actions (Continued)

Social/ relational	Misunderstandings, lack of feedback, exclusion from decisions, cultural clashes, gossip	Withdrawal, tension in meetings, reluctance to speak up, more "work-to-rule" behaviour	Team climate pulse score (e.g., 5-item survey monthly: baseline 4.0/5, disrupted 3.0/5); participation rate in meetings (% of members contributing)	Regular team check-ins, clear ground rules for discussion, cross- cultural awareness, role clarity, structured feedback routines	Mediated conversation, "reset" meeting to clarify expectations, agreeing on next steps and shared goals, temporarily separating high-conflict tasks/people
Professional/ organisational	Rapid change, unclear priorities, shifting key performance indicators, restructure, poor role design	Confused effort, duplicated work, missed handovers, project delays, staff turnover	Priority changes per month (baseline 1–2 major changes, disrupted 6+); turnover/ transfer rate in team per year (baseline 5–10%, spike to 25%)	Clear strategy and cascading goals, stable core routines, transparent communication of changes, realistic workload planning	Re-negotiating priorities with management, pausing non-critical projects, clarifying decision rights, short "re-planning" huddle with key stakeholders

Positive self-talk and cognitive reframing play a crucial role in how setbacks are handled. When a project doesn't unfold as expected, it helps to see it as a challenge to learn from rather than a personal failure. That shift in perspective, what many call an "optimism habit", creates space for growth instead of self-doubt. It's about turning obstacles into lessons, not verdicts. Nelson Mandela captured this perfectly: *I never lose. I either win or learn.* Embracing that mindset builds resilience and steadiness, even when uncertainty looms large. Over time, this approach transforms setbacks from roadblocks into stepping stones towards progress.

Physical movement plays a vital role too. A short workout in the morning lifts my energy and clears mental fog. Exercise triggers chemical changes that ease tension and support recovery. On days when stress builds, movement helps me reset and reconnect with my sense of control.

These practices are small, yet they carry weight. They can prepare us to face emotional disruptions with strength and grace, like tuning an instrument before stepping onto the stage.

- *Reactive Strategies:* When emotional jolts strike, we need to reach for tools that help us reset quickly, like grabbing a life vest in rough waters. Over time, we need to learn that having simple, reliable techniques ready makes all the difference. Deep breathing, especially box breathing (inhaling for four seconds, holding, then exhaling slowly), helps calm the storm inside. It's like flipping a switch that quiets the fight-or-flight response and brings me back to the centre.

  Sometimes, writing things down, just a few lines in a journal, gives shape to emotions that otherwise feel overwhelming. Putting thoughts on paper turns vague feelings into something tangible, something that can be understood rather than carried like an invisible weight. Other times, talking it through with someone trusted brings clarity and perspective that silence can't offer. These small acts of reflection create space to process rather than suppress. Even something as simple as stepping outside for a short walk can shift energy, clear mental fog, and remind us that movement often sparks momentum. In those moments, the mind feels lighter, and the day becomes easier to navigate.

In those moments, we should ask ourselves: *What do I need right now?* Maybe it's a breath, a kind word to myself, or a few minutes away from the noise. That pause creates space, space to choose a response instead of reacting automatically. If a strong emotion stays, we should remind ourselves: *It's okay to feel this. I've faced hard things before.* That gentle self-compassion softens the edges and restores clarity.

And we should never underestimate the power of connection. A quick message to a friend or a quiet chat with a colleague often dissolves anxiety faster than any solo strategy. Research consistently shows that social support acts like emotional armour, one of the strongest protective factors in resilience [5]. For example, a study published in *Health Psychology* found that people who received emotional support from close friends during stressful events had lower cortisol levels and faster cardiovascular recovery. Another practical example comes from workplace settings: employees who regularly check in with supportive colleagues or mentors report higher job satisfaction, better emotional regulation, and reduced burnout. Even something as simple as a five-minute conversation with a trusted friend can shift your emotional state, offering perspective and calm.

In my own experience as a safety lead in different industrial sectors, reaching out to someone, even just to say, *I'm having a rough day*, often dissolves tension more effectively than any solo strategy. It's like stepping out of a storm and finding shelter. Whether through a quick message, a walk with a friend, or a quiet moment with someone who listens, social connection transforms how we carry stress.

EXAMPLE SCENARIO

Imagine receiving an unexpected reprimand at work. The first wave of emotion, anger, frustration, maybe even shame, hits hard. Instead of letting it spiral, pause for two minutes of mindful breathing. Name the feeling: *I feel frustrated.* That simple act of labelling creates distance and clarity. Then, remind yourself: one critique doesn't define your worth or your abilities. A quick stretch or a short walk can reset your energy before you return to the conversation. Approaching the issue calmly with your manager becomes easier when the emotion is acknowledged rather than resisted. By processing it instead of suppressing it, the impact shortens, and productivity stays intact.

MENTAL FOCUS (MENTAL/ COGNITIVE DISRUPTIONS)

Mental disruptions often feel like static in the brain, an invisible fog that blurs clarity and drains energy. They show up as anxiety and worry, especially about problems that haven't even happened yet. They creep in through cognitive clutter from multitasking, intrusive thoughts that hijack focus, and the relentless flood of information we absorb every day. In today's hyper-connected world, true mental silence is rare. There are days when the mind is juggling emails, replaying a difficult conversation, and racing to meet a deadline, all at once. The result? Slower thinking, more mistakes, and a quiet sense of exhaustion that builds beneath the surface. It's not dramatic, but it's draining, and it reminds us how vital mental clarity has become in a world that never stops talking.

Research shows that switching between tasks creates a measurable "switching cost" [6]. Each time we shift focus, the brain must reorient itself, which adds delays and increases the likelihood of errors. It's like trying to read a book while someone keeps changing the page. Stress and negative thinking also consume working memory, the mental workspace we rely on for reasoning and decision-making. When that space is filled with worry, there's little room left for creativity or problem-solving.

One strategy that consistently helps is single-tasking with intention. Blocking time for deep work and silencing notifications creates a pocket of uninterrupted focus that feels like mental fuel. Even 25 minutes of pure concentration can restore clarity and momentum. Another practice is mental decluttering, writing down lingering thoughts before starting a task so they don't hover in the background like static. It's a way of clearing the runway before take-off, ensuring the mind is ready for take-off rather than weighed down by distractions. These small habits don't just improve productivity; they protect mental energy in a world that constantly pulls attention in every direction.

As Bruce Lee once said, *The successful warrior is the average man, with laser-like focus.* That kind of focus isn't automatic; it's built through habits that protect our cognitive space. Whether it's a short meditation, a walk to reset, or simply saying "no" to one more tab opened, these small acts help preserve mental clarity.

- *Proactive Strategies:* Promote cognitive habits which favour focus, like sharpening a blade before stepping into the day. We should've learned that mental clarity doesn't arrive by accident; it's something we must create, protect, and renew. One of the most powerful habits we should've adopted is single-tasking. When we dedicate a block of time to one high-priority task, we feel more present, more capable. The Pomodoro Technique, working in 25-minute focused bursts followed by short breaks, respects the brain's natural rhythm. It's like giving your mind a sprint, then a breath.

To keep our thoughts organised, we offload them. We should write to-do lists, sketch out time blocks, or use digital notes. This simple act frees up mental space and eases the pressure of "remembering everything". It's like clearing clutter from a desk before starting work, suddenly, there's room to think.

When my mind starts looping, replaying conversations or worrying about what's next, I turn to mindfulness. Even five minutes of quiet breathing or a short meditation helps me silence the mental noise. It's not about emptying the mind, but about noticing the chaos and choosing not to follow it. As Marcus Aurelius wrote, *You have power over your mind, not outside events. Realize this, and you will find strength.*

Another habit that makes a difference is pre-planning. When a difficult conversation or decision is on the horizon, rehearsing how to respond can ease the tension. Visualising the moment, preparing the words, and imagining staying calm creates a sense of control before the situation even arrives. It's like laying down tracks before the train comes, when it does, you're ready. This simple mental rehearsal prevents anxious loops from taking over and turns uncertainty into something manageable. Instead of reacting in the heat of the moment, you respond with clarity and confidence.

These habits don't just improve focus, they protect it. They help us show up with clarity, even when the world feels loud.

- *Reactive Strategies:* When the mind starts to wander or spiral into stress, quick mental resets can feel like gently steering a drifting boat back on course. One of the most effective tools is the 5-4-3-2-1 grounding technique: name five things you can see, four you can feel, three you can hear, two you can smell, and one you can taste. This simple exercise pulls attention out of anxious loops and anchors it in the present moment. It's like switching the channel from noise to clarity, a reminder that calm often begins with awareness.

  Sometimes, standing up and stretching is enough to break the cycle of overthinking and remind you that you're present and in control. When multitasking creeps in, jumping between tabs or reaching for the phone, a quick pause helps. Close one distraction, mute notifications, and gently return attention to the task at hand. These small actions clear mental fog and restore focus. Reducing digital clutter amplifies this effect. Removing unused or distracting apps like Facebook, Instagram, or TikTok prevents endless scrolling and wasted time. Keeping Facebook only on a computer, for example, eliminates constant phone notifications. Limiting news consumption also matters, if something truly important happens, it will reach you without hours of headline scanning. Protecting attention in these ways keeps the mind clear, focused, and anchored on what matters most.

We also need to embrace mental pauses throughout the day. Even 30 seconds of focused breathing or a quiet stretch halfway through a task can reset our focus. It's not about doing more; it's about doing more with presence.

When negative thoughts surface, like *I'm behind* or *I'm messing this up*, the first step is to label them: *That's a stress-thought.* Naming the thought creates distance and reduces its grip. The next step is to challenge it: *I've handled tough deadlines before.* This simple act of cognitive reappraisal shifts the mind from panic to perspective. Instead of spiralling, you regain control and remind yourself that setbacks are temporary, not defining. Over time, this habit turns self-doubt into resilience.

As the philosopher William James once said, *The greatest weapon against stress is our ability to choose one thought over another.* That choice begins with awareness and grows stronger with practice.

EXAMPLE SCENARIO

When negative thoughts creep in, those familiar lines like "I'm behind" or "I'm messing this up", the first instinct is often to believe them. But pausing to label them changes everything: *That's a stress-thought.* Naming the thought creates a small but powerful distance between you and the emotion. It reminds you that thoughts are not facts, they're mental events passing through. The next step is to challenge them with evidence: *I've handled tough deadlines before. I've solved harder problems.* This isn't about blind positivity; it's about reframing the narrative, so it reflects reality rather than fear. Psychologists call this cognitive reappraisal, and it works because it shifts the brain from panic mode into problem-solving mode. Instead of spiralling into self-doubt, you regain perspective and control. Over time, this habit becomes a mental muscle, one that turns stress into resilience and keeps you moving forward even when pressure mounts.

PHYSICAL WELL-BEING (PHYSICAL DISRUPTIONS)

When the body feels depleted, whether from poor sleep, chronic stress, or skipping meals, everything else becomes harder. Physical disruptions like fatigue, illness, or injury act as silent brakes on performance, slowing progress in ways that aren't always obvious at first. Even with the best intentions to push through, the signs appear: focus slips, patience thins, and decisions become slower and less precise. It's like trying to run a marathon with sandbags tied to your legs, possible, but exhausting and unsustainable. The truth is that mental clarity and emotional resilience are deeply tied to physical well-being. When the foundation cracks, the entire structure wobbles. That's why tending to the basics, sleep, nutrition, movement, often feels less like self-care and more like strategic maintenance for the mind.

Sleep is a cornerstone of cognitive health. Multiple studies confirm that fragmented or insufficient sleep directly impairs mental performance [7–8]. The effects are hard to ignore, after a restless night, reaction times slow, memory falters, and even simple tasks feel like uphill climbs. Fatigue doesn't just make you tired; it rewires how the brain processes information, turning clarity into fog. Chronic stress adds another layer of strain. Elevated cortisol levels disrupt hormonal balance and drain energy reserves, leaving the body in a constant state of alert. When the body is under pressure, the mind struggles to stay sharp. Focus slips, patience thins, and decision-making becomes slower and less precise. It's a powerful reminder that mental resilience isn't just about mindset, it rests on the foundation of physical well-being. Without that base, everything else wobbles.

That's why we treat physical care as a foundation, not a luxury. We need to have learned to prioritise sleep like a meeting with our future self. We need to keep a wind-down routine, limit screens before bed, and aim for consistency. We also pay attention to nutrition, eating regularly and choosing foods that fuel rather than fog our brain. Movement matters too. Even a short walk or stretch between tasks helps us reset and recharge.

As Jim Rohn wisely said, *Take care of your body. It's the only place you have to live.* When we care for our physical well-being, we strengthen every other part of our performance system. The body fuels the mind, and when it's supported, we're far better equipped to handle life's disruptions.

- *Proactive Strategies:* Invest in baseline physical health, the foundation of resilience and sustained performance. When we care for our bodies, we create the conditions for our minds to function at their best. First, we prioritise sleep hygiene. Aiming for seven to nine hours of quality sleep each night helps restore energy and sharpen focus. This includes a consistent bedtime, a cool and dark room, and reducing screen time before sleep. Adequate sleep refreshes neural connections and prepares us to handle the demands of the day.

Second, we make regular exercise part of our routine. Research shows that physical activity strengthens both cognitive and emotional resilience [9]. Even moderate movement such as walking, cycling, or yoga releases endorphins and stress-buffering hormones. Over time, it builds a stronger brain, often referred to as cognitive reserve, which helps us recover more quickly from mental and emotional disruptions.

Third, we maintain basic nutrition and hydration. Low blood sugar or dehydration can mimic stress in the brain, leading to irritability and fatigue. Carrying healthy snacks, drinking water regularly, and avoiding extreme caffeine or sugar swings helps us stay balanced and alert.

As Hippocrates said, *let food be thy medicine and medicine be thy food.* When we nourish our bodies, we strengthen every other part of our performance system. Physical well-being is not separate from mental clarity. It is the engine that powers it.

- *Reactive Strategies:* When a physical disruption appears, such as sudden fatigue or discomfort, we respond quickly to prevent it from growing into a barrier. These signals from the body deserve attention. If we feel overwhelmed by tiredness, a short power nap or even lying down for 10 to 20 minutes can restore clarity and energy. Light stretches or mobility exercises at our desk help increase blood flow and improve alertness.

  When we notice early signs of illness, such as a headache or stomach discomfort, taking one small step can make a difference. Drinking water, using appropriate medicine, or stepping away to rest for a few minutes helps us stay functional. These actions are simple, yet they protect our ability to think clearly and stay focused.

  During stressful moments, a quick body scan, mentally checking in on areas of tension or discomfort, can reveal what we need. Sometimes it is a breath, sometimes movement, sometimes a short pause. These small responses act like a reset button, helping us stay steady throughout the day.

  As Leonardo da Vinci said, *The human foot is a masterpiece of engineering and a work of art.* Our bodies are designed to support us, and when we listen to them with care, we strengthen our resilience from the inside out.

EXAMPLE SCENARIO

You're running on an hour's sleep after a late night. Mid-morning, you start nodding off. Instead of pushing through for hours, you stand, take five deep breaths, drink a glass of water, and walk briskly around the block. Even this five-minute reset raises your alertness. Later, you decide on a short evening nap and plan to go to bed an hour earlier that night, preventing a cycle of chronic sleep loss.

DIGITAL DISTRACTIONS (TECHNOLOGY DISRUPTIONS)

In our always-connected world, technology serves both as a powerful tool and a frequent source of disruption. Smartphone pings, endless email threads, and social media alerts constantly pull our attention in different directions. These digital stimuli fragment our focus and make it harder to stay present. Cognitive science shows that every notification forces the brain to switch tasks, which slows us down and increases the chance of errors.

Social media is especially designed to keep us engaged. Each like, message, or update triggers a dopamine loop, creating a cycle of variable reinforcement that makes disengagement difficult. Over time, this constant engagement shortens our attention spans and weakens our ability to do deep, focused work. One review found that employees who frequently use social media during work hours experience lower task completion rates, reduced creativity, and higher stress levels.

We have all felt this. A quick glance at a message can turn into 20 minutes of scrolling. A single email can lead to a cascade of distractions. These moments may seem small, but they accumulate and drain our mental energy.

To protect our focus, we create boundaries around technology. We silence notifications during deep work, schedule specific times to check messages, and keep our phones out of reach when we need to concentrate. These habits help us reclaim control over our attention.

As Cal Newport, author of *Deep Work*, said, *Clarity about what matters provides clarity about what does not*. When we choose to engage with technology intentionally, we strengthen our ability to focus, create, and perform at our best.

- *Proactive Strategies:* We establish digital boundaries, so technology supports us, rather than controls us. In a world filled with constant alerts and notifications, we choose when and how to engage. One of the first steps is managing notifications. We turn off non-essential alerts on our phones and computers, especially during focused work or personal time. Tools like app blockers or do not disturb settings help us protect our attention.

  We also create structured routines. For example, we check email or social media only at scheduled times, such as once mid-morning and once mid-afternoon. This prevents us from reacting to every ping and keeps our minds clear. The Pomodoro Technique, mentioned earlier, supports this approach. During each focus interval, we keep screens silent and distractions away.

  A daily tech fast can also help. We might choose to avoid phones during meals or for the first hour after waking. These small habits reduce the chance of falling into a scroll or email spiral and help us stay present with ourselves and others.

  As Mahatma Gandhi said, *Action expresses priorities.* When we set boundaries around technology, we show that our time, focus, and well-being matter. These choices allow us to use digital tools with intention and protect the space we need for deep work and meaningful connection.
- *Reactive Strategies:* When a digital distraction appears, we respond with intention to protect our focus. If a surprise message pops up, we pause before clicking. We ask ourselves, is this urgent or can it wait 20 minutes? If it can wait, we close the notification and return to our task. This small decision helps us stay in control of our attention.

  When the distraction is more open-ended, such as scrolling through social media, we use a countermeasure. We might stand up, take a drink of water, or do a short breathing exercise. These actions help reset the mind and break the loop. Some people use the two-minute rule. If opening an app or browser takes more than two minutes, we do a quick check-in on our task instead. This helps us reconnect with what matters.

  Visual cues also support focus. A sticky note on the desk or a desktop wallpaper with a reminder like "Finish project before lunch" can snap our attention back. These small prompts act like mental anchors.

  As Seneca said, *It is not that we have a short time to live, but that we waste much of it.* When we respond to digital distractions with clarity and purpose, we reclaim our time and strengthen our ability to stay present.

EXAMPLE SCENARIO

You install a distraction blocker that mutes all notifications during your morning work block. Later, your phone buzzes – a friend texting about weekend plans. Normally you might interrupt your flow to reply. Today, you take a breath and note it: *I'll reply in 30 minutes.* You stay on task, then after finishing your work task, you give yourself a short break to catch up on messages. With proactively scheduled tech checks and pausing before responding, you preserve deep focus and avoid constant jolts.

SOCIAL ENVIRONMENT (SOCIAL DISRUPTIONS)

We are social creatures, and our relationships shape how we think, feel, and perform. Supportive interactions, a kind word from a colleague, a moment of care from a partner, can act as emotional buffers, helping us stay steady during stress. These moments of connection remind us that we are not alone, and they often restore clarity and motivation.

On the other hand, social disruptions can be deeply destabilising. A harsh comment from a manager, tension with a teammate, or conflict at home can trigger emotional reactions that linger. Feelings of anger, sadness, or rejection can cloud our thinking and drain our energy. These disruptions do not stay confined to the moment, they echo through our day and affect how we show up in every task.

Research on resilience highlights the importance of social support. People with strong networks tend to recover more quickly from adversity. Whether through family, friends, or trusted peers, these connections act like emotional scaffolding, helping us rebuild when things fall apart. Our social context can either amplify stress or absorb it.

We can strengthen this protective layer by nurturing relationships that are built on trust, empathy, and mutual respect. A regular check-in with a friend, a moment of appreciation for a colleague, or simply listening with presence can reinforce these bonds. These small acts create a buffer against disruption.

As Helen Keller said, *Alone we can do so little; together we can do so much.* When we invest in healthy social environments, we build a foundation that supports resilience, focus, and well-being.

- *Proactive Strategies:* We nurture positive connections before trouble arrives. Supportive relationships are a vital part of resilience. We seek out mentors, friends, and colleagues who are encouraging and solution-oriented. These people form a small support team we can lean on during difficult times. Strong friendships and collaborative relationships act like a reserve tank of emotional strength. In the workplace, this might mean finding a peer or coach who helps us navigate stress. At home, it means communicating openly about challenges and asking for help when needed.

We also set boundaries with difficult individuals. If a colleague is frequently negative or interruptive, we limit conversations or schedule specific times to engage. If a family member's habits disrupt our focus, such as constant complaints or requests, we practice saying, we need a moment, or suggest discussing issues at a designated time. Choosing when and how we engage, we reduce unwanted social shocks and protect our energy.

Importantly, many people refer to a category known as toxic individuals. This includes those who display traits from the Dark Triad: narcissism, Machiavellianism, and psychopathy. Learning to identify these patterns early is essential. Our response depends on the nature and depth of the relationship. For example, if someone shows narcissistic tendencies, we might set firm boundaries in casual interactions or seek professional support in more complex situations. Recognising these traits and adapting our strategy helps preserve emotional energy and maintain resilience.

As Eleanor Roosevelt said, *No one can make you feel inferior without your consent.* When we choose our social connections with care and respond to disruptions with clarity, we build a stronger foundation for focus, well-being, and growth.

- *Reactive Strategies:* When a social disruption occurs, we respond with care and intention to prevent it from spreading through the rest of our day. Whether it is an argument, an unexpected change in plans, or a moment of tension, the first step is often to pause and breathe. Even one or two deep breaths can help us stay grounded and avoid reacting in anger.

  Active listening is one of the most underrated skills in conflict resolution. Many disagreements escalate simply because someone feels unheard. Paraphrasing what the other person is expressing, such as saying, *I hear that you feel overwhelmed*, can diffuse tension and create space for understanding. It signals empathy and opens the door to resolution rather than resistance. When emotions feel heavy, seeking an outside perspective helps. A brief walk with a colleague or a short conversation with a trusted friend can act as a release valve, allowing you to process what you're carrying instead of bottling it up. These small, intentional actions remind us that clarity often begins with connection, listening to others and letting ourselves be heard.

When a social situation leaves us feeling hurt, we use empathy-based self-talk. Reminding ourselves, this person is also human and may be having a difficult day, can soften the emotional sting and restore perspective. These small mental shifts help us stay steady.

After the disruption, we reconnect with a source of support or comfort. This might be calling a friend, spending time with a pet, or doing something kind for ourselves. These actions help contain the emotional impact and prevent it from affecting our focus and well-being.

As the Dalai Lama said, *Be kind whenever possible. It is always possible.* When we respond to social disruptions with compassion and clarity, we protect our emotional energy and strengthen our resilience.

EXAMPLE SCENARIO

During a meeting, a colleague unexpectedly criticises your idea. You feel defensive and your heart races. Instead of snapping back, you quietly note your feelings (*I'm hurt/angry*) and take a slow breath. You then calmly respond, "Thanks for the feedback; let me think on it", buying a little time. After the meeting, you send that colleague a quick clarifying message (to ensure you understood correctly) and share the situation with a friend later for support. Pausing and then validating rather than venting, you preserve the working relationship and prevent a larger emotional fallout.

PROFESSIONAL CONTEXT (WORKPLACE DISRUPTIONS)

Professional disruptions can feel like storms crashing against our internal stability. These include sudden crises, shifting roles, unreasonable demands, or chaotic environments. When priorities change without warning, when deadlines appear overnight, or when meetings pile up without purpose, we often feel scattered and overwhelmed. These moments challenge our ability to stay focused and composed.

A toxic work culture can create ongoing disruption. Lack of support, bullying, or excessive workloads do more than frustrate us, they erode our motivation and well-being. Occupational psychology has shown that toxic leadership and chronic overload are linked to lower job satisfaction and higher rates of burnout. When the environment itself becomes unpredictable or hostile, it injects stress into every part of our system.

We respond by creating structure and clarity wherever possible. We clarify expectations, ask for realistic timelines, and prioritise tasks with our team. We also protect time for deep work, using tools like time blocking or scheduled focus sessions. These habits help us stay fastened when the workplace feels unstable.

We also advocate for ourselves. If demands become unreasonable, we speak up respectfully. If the culture feels harmful, we seek allies, mentors, or external support. These actions are not signs of weakness; they are signs of resilience.

As Viktor Frankl said, *When we are no longer able to change a situation, we are challenged to change ourselves.* In the face of workplace disruption, we choose to respond with clarity, courage, and care.

- *Proactive Strategies:* Wherever possible, we shape our professional environment to reduce surprises and create clarity. This begins with setting expectations. We ask for clear project scopes and realistic deadlines, so we are not caught off guard by shifting goals. We prioritise tasks collaboratively with our team or manager to avoid last-minute rushes. Breaking large projects into smaller milestones and scheduling regular check-ins helps prevent sudden crises and keeps progress visible.

  Protecting time is essential for maintaining focus and reducing mental clutter. One effective approach is time-blocking, scheduling dedicated blocks for deep work and marking them as "busy" so others know you're unavailable. These boundaries signal that focus matters. If the workplace allows, supporting initiatives like deep workdays, where meetings are minimised and concentration is encouraged, can make a huge difference. Clear communication reinforces these habits. A simple message such as, *I'll respond to non-urgent emails by noon each day*, sets expectations and reduces interruptions. These small steps create breathing room for meaningful work and prevent the constant pull of reactive tasks. In a world of endless notifications and competing priorities, protecting time isn't just a productivity tactic, it's a way to safeguard mental clarity and creative energy.

  Feedback is another essential part of a resilient work environment. We actively seek input from trusted colleagues and our direct manager. Constructive feedback is a necessity, not a luxury. It is the responsibility of leaders to provide it regularly and meaningfully. If feedback is missing, it may reflect a breakdown in communication or a lack of investment in our growth. Both situations deserve attention.

Improvement requires engagement from all stakeholders. Feedback should flow both ways and be part of the culture, not limited to formal reviews. When feedback loops are open and honest, they become powerful tools for resilience, clarity, and continuous development.

As Ken Blanchard said, *Feedback is the breakfast of champions.* When we create environments that support clarity and growth, we give ourselves and others the best chance to thrive.

- *Reactive Strategies:* When a professional disruption hits, as it inevitably will, the key is to tackle it systematically rather than reactively. Start with assessment and organisation. Write down the immediate tasks the disruption has created and choose one to focus on first. Apply quick triage: what absolutely must be done in the next hour, what can wait until tomorrow, and what belongs in the next week? If possible, delegate or negotiate: *I can handle this by tomorrow; for the rest, I may need help.* Next, apply stress-control techniques so panic doesn't cloud judgement. A few deep breaths, a short walk, or reframing the situation can restore clarity. Then, move to restoring order. Once the immediate crisis is stabilised, rebuild your routine. For example, if a meeting ran long and derailed your schedule, quickly reschedule affected tasks to regain structure. Finally, debrief, even briefly. Ask: *What caused this disruption, and how can I prevent it next time?* Then take one small action, like updating a process or communicating a new boundary, to reduce the chance of repeat. This approach turns chaos into a controlled response and keeps momentum intact.

EXAMPLE SCENARIO

Late in the day, your manager tells you a report must be doubled in length for an early-morning meeting tomorrow. Your first response might be stress, but you take a quick breath and ask questions: "Which sections need more detail?" This narrows the scope. You delegate some data gathering to a teammate, then outline and write key points. After you submit the expanded report, you reflect: *If I have to do this again, I'll ask earlier and plan time earlier.* You email your manager tomorrow morning confirming the new length in advance. Reacting calmly and then adjusting your workflow and communication, you minimise the disruption's damage and prepare to avoid it next time.

QUICK PRACTICE: TACKLING YOUR OWN DISRUPTIONS

Step 1: Identify a Frequent Disruption.

Think of a disruption which often knocks you off balance – it could be an emotional upset, a digital distraction, a recurring task crisis, or something else. Write one sentence describing it.

My Disruption: ________________________________

(Example: I constantly check my phone during work and lose focus.)

Step 2: Determine Its Category.

Label the type of disruption (emotional, mental, physical, digital, social, or professional). Consider what typically triggers it.

Type of Disruption: ________________________________

(Example: Digital – I get notifications and breaks in concentration)

Step 3: Choose a Proactive Change.

Based on the category, pick one preventive action you can try *before* the disruption happens. Write a simple plan or rule.

Preventive Strategy: ________________________________

(Example: I will disable non-urgent notifications during my focus hour.)

Step 4: Choose a Coping Move.

Select one quick coping technique to use *in the moment* if this disruption occurs. This should be something simple you can do immediately.

Coping Strategy: ________________________________

(Example: If I feel an urge to check my phone, I'll take 5 deep breaths first.)

Try this out this week. Next time your chosen disruption appears, follow your new strategy and note what happens. Over time, small adjustments like these – cutting off a distraction before it starts or pausing instead of reacting – will add up to stronger, smoother performance. Progress happens one step at a time, and each intentional change helps your personal system run more reliably and resiliently.

REFERENCES

1. Nandyala, R., 2025. *Analog brain*. BlueRose Publishers Pvt Ltd.
2. Lee, H.P., Chiang, Y.S., Gao, L., Yang, S., Winter, P. and Das, S., 2025. Purpose mode: Reducing distraction through toggling attention capture damaging patterns on social media web sites. *ACM Transactions on Computer-Human Interaction*, 32(1), pp.1–41.
3. Wei, X., Ma, J., Liu, S., Li, S., Shi, S., Guo, X. and Liu, Z., 2025. The effects of sleep deprivation on risky decision making. *Psychonomic Bulletin & Review*, 32(1), pp.80–96.
4. Hall-Renn, K.E., 2007. Mindful journeys: Embracing the present with non-judgmental awareness. *Journal of Creativity in Mental Health*, 2(2), pp.3–16.
5. Hazan-Liran, B. and Walter, O., 2025. Mediating role of social support in relations between psychological capital and subjective well-being among IDF soldiers during conflict: Insights from the 2023 Gaza war. *Israel Journal of Health Policy Research*, 14(1), pp.1–14.
6. Yang, Y., Chuang, Y.T., Wu, C.H. and Chen, Y.J., 2025. *Is switching cost always a first-mover advantage for the technology leader?* Available at SSRN 5143033.
7. Benkirane, O., Simor, P., Mairesse, O. and Peigneux, P., 2024. Sleep fragmentation modulates the neurophysiological correlates of cognitive fatigue. *Clocks & Sleep*, 6(4), pp.602–618.
8. Georgeson, T., Atkins, L., Zahnleiter, A., Terrill, P.I., Eeles, E., Coulson, E.J. and Szollosi, I., 2025. Sleep fragmentation and hypoxaemia as key indicators of cognitive impairment in patients with obstructive sleep apnoea. *Sleep Science and Practice*, 9(1), p.2.
9. Zhu, D. and Li, Y., 2024. Boosting happiness in left-behind children: Unveiling the power of physical activity through cognitive reappraisal and psychological resilience. *PLOS One*, 19(10), p.e0309291.

The First Breath After Impact *Stabilising Before Recovery*

3

Out of suffering have emerged the strongest souls; the most massive characters are seared with scars.

Kahlil Gibran

OPENING REFLECTION

There is a moment in every disruption where the air feels heavy, where silence presses in, and where time itself seems to slow. Performance had been smooth, almost automatic, then came the shock. The shock. The unexpected wave that knocked us off course. In those first moments, our ability to *absorb* the impact determines whether we shatter or bend.

Resilience scholars describe resilience as unfolding across stages: functionality, disruption, absorption, adaptation, recovery, and learning. In this framework, the absorption period is the crucial bridge between being hit and beginning to rebuild. It is when the system, whether an engineered structure, a company, or a human being, faces the immediate turbulence of the shock and decides, through its inherent design, whether to hold together or collapse.

For humans, this does not look like a mechanical process. It is visceral, biological, emotional, and deeply personal. The body surges with adrenaline. The heart pounds. The amygdala, the brain's alarm bell, hijacks rational thought. And yet, within this storm lies an opportunity: to stabilise, to contain the damage, and to absorb the impact long enough to regain clarity.

DOI: 10.1201/9781003789895-4

This chapter explores how we can manage the absorption period, drawing from psychology, neuroscience, resilience engineering, and our own lived experiences. It is all about learning to ride out the first wave, protecting core functionality while the storm passes, and buying the time we need for adaptation and recovery to begin.

THE SCIENCE OF ABSORPTION

Stress Response and the Body

When a shock occurs, whether a job loss, a harsh critique, a medical scare, or even a sudden digital disruption, the nervous system launches into fight-or-flight mode. The sympathetic nervous system floods the body with adrenaline and cortisol. Muscles tighten, pupils dilate, breathing quickens. The brain diverts resources away from rational analysis and into the survival response.

This is what neuroscientists call an "amygdala hijack": the emotional centre of the brain overrides the prefrontal cortex which normally manages planning, reasoning, and decision-making [1]. Psychologist Daniel Goleman coined the phrase to describe how sudden shocks trigger emotional reactions before logic has a chance to intervene [2].

From an evolutionary standpoint, this reaction is protective. If a predator jumped from the bushes, hesitation could mean death. But in modern disruptions, a tough email from a manager, a financial setback, or a health scare, the same ancient response often works against us. We freeze, lash out, or panic. Our PF dips sharply.

The absorption phase is the period during which this stress cascade is active but can be regulated. Scientists studying resilience in both engineering and psychology agree: the sooner a system can contain the initial shock, the less long-term damage occurs [3–4]. In human terms, absorption is actually grounding ourselves, avoiding rash actions, and stabilising enough to remain functional.

ABSORPTION IN SYSTEMS THEORY

In resilience engineering, systems are designed with buffers, redundancies, and dampers which allow them to absorb shocks. A bridge sways with the wind instead of snapping. A resilient information technology (IT) network reroutes data when a server crashes. A resilient supply chain has alternate suppliers when one is disrupted.

Humans, too, have buffers, resources that help us absorb life's shocks and maintain balance. These include social support, coping habits, emotional regulation skills, routines, and self-care practices. When we invest in these areas, we increase our capacity to handle stress and uncertainty. But when these buffers are weak or missing, even small disruptions can feel overwhelming.

Now, let's talk about reading, especially self-development, spiritual, and psychological books. When was the last time you truly engaged with one of these genres? Many people feel sceptical about their practical value. It's true: we often forget the content, or struggle to apply it consistently in real life. Take popular titles like *Atomic Habits* [5] or *The Psychology of Money* [6]. These books promote powerful ideas, like building habits or creating financial independence beyond the 9-to-5 grind. *The Psychology of Money*, for example, has sold over 9 million copies by now. But how many readers have actually transformed their lives or started their own businesses because of it? Probably less than 1%.

Still, that doesn't mean reading is pointless.

Reading is a catalyst. It may not instantly change your life, but it plants seeds, ideas, perspectives, and motivation. It boosts your mindset, expands your emotional vocabulary, and helps you reflect on your values. Even if only a fraction of the content sticks, it can shape your decisions, your resilience, and your personal growth over time. The real power lies not just in the information, but in how it shifts your thinking and opens new possibilities.

The National Academies of Sciences defines resilience as *the ability to absorb, adapt to, and recover from a disruptive event without significant deviation from expected performance.* Absorption, then, is not passive endurance. It is active containment, holding the line until recovery mechanisms can take over.

STORY: THE LAYOFF EMAIL

The day began like any other, until the inbox revealed a subject line that froze time: "Restructuring Update – Your Role". The email opened with clinical precision: *Your role has been made redundant effective immediately. HR will contact you for next steps.*

For a few seconds, breathing felt impossible. A tightness gripped the chest, palms dampened, and thoughts spiralled: *What will the family think? How will rent be paid? Does this define worth?*

That was the absorption period, the first hour when stress chemicals hijack the body and the mind catastrophises: *This isn't just a job loss; it's the end of everything.* The apartment became a pacing track. The email refreshed again and again, as if words might magically change. An angry response almost took shape. Three tearful calls almost happened.

Then, a line from an old journal surfaced like a lifeline: "In the absorption phase, don't make permanent decisions in temporary storms". So, the storm was allowed to pass. No fixing, no forcing clarity. Just sitting on the couch, a hand placed gently on the chest, breathing slowly, deliberately, with intention.

Each breath served as an anchor, a quiet reminder of presence and resilience. The storm within was acknowledged without judgement: This is a shock. It feels terrible. But it is not the end. There was no attempt to suppress discomfort or rush towards solutions. Instead, space was given for absorption, to let emotions settle, to allow intensity to pass through without amplifying it with impulsive reactions. Permission was granted to simply be, to feel, and to wait for clarity to return.

In that moment, the power of pausing became clear, the strength of creating space between the trigger and the response. Within that space lives resilience. Within that space, wisdom takes root. Absorb first. Respond later. This is not weakness; it is strategy. It is self-respect. It is the quiet strength of someone who chooses to feel fully yet act wisely.

By the end of the day, nothing had shifted on the surface. Unemployment remained. Uncertainty lingered. Questions about the future stayed unanswered. But internally, something significant had occurred: no bridges were burned, no impulsive decisions made, no collapse under the weight of panic. The shock had been contained.

That containment was never about suppressing emotion; it was about holding space for it. Discomfort was allowed to exist without dictating actions. In doing so, something vital was protected: the ability to respond with clarity rather than chaos. That evening, quiet reflection revealed a shift. Containment had created room for something else to emerge: adaptability. In the days that followed, thoughts became clearer. Connections were made thoughtfully. New possibilities were explored with a steadier mind. Progress wasn't about rushing to fix everything; it was about moving forward, one grounded step at a time.

Sometimes, the most powerful progress isn't loud or dramatic. It's the quiet decision to stay present, to not unravel, and to trust that recovery begins with resilience.

DOMAINS OF ABSORPTION

Absorbing shocks is a complex process that goes beyond emotional regulation. It touches multiple areas of our lives, each with its own vulnerabilities and strengths. Just as Chapter 2 explored the various sources of disruption, whether personal, societal, or technological, this chapter focuses on how we can navigate the absorption phase across six key domains: emotional, mental, physical, digital, social, and professional.

Each domain plays a distinct role in how we process and respond to unexpected events. Emotionally, we need tools to manage intense feelings like fear, grief, or anger. Mentally, we rely on clarity, focus, and cognitive flexibility to make sense of what's happening. Physically, our bodies carry stress, so movement, rest, and nourishment become essential. Digitally, we must be mindful of the information we consume and the boundaries we set online. Socially, our relationships can either support or strain us during difficult times. And professionally, our work environment and career stability can be deeply affected by shocks, requiring adaptability and strategic thinking.

Understanding these domains helps us build a more holistic approach to resilience. Instead of reacting impulsively or feeling overwhelmed, we learn to absorb the impact thoughtfully, giving ourselves the space to recover and respond with intention.

Emotional Absorption

Emotions that follow a shock are much like aftershocks that follow an earthquake. They don't arrive in a neat, predictable pattern. Instead, they surge unexpectedly, sometimes hours or days later, threatening to destabilise us even further. These emotional waves can feel just as intense as the initial impact, catching us off guard and making it difficult to think clearly or act wisely.

From a scientific perspective, emotional shocks activate key regions of the brain, particularly the amygdala and the broader limbic system. These areas are responsible for processing fear, threat, and emotional intensity. When triggered, they narrow our focus, heighten our sensitivity to negative stimuli, and reduce our access to rational thought. This is why, in moments of distress, we often feel overwhelmed, reactive, and unable to see the bigger picture.

However, neuroscience also offers a powerful tool for managing this response: affect labelling [7]. This is the simple act of naming what we're feeling, saying to ourselves, "I feel angry", or "I feel scared". Research shows that when we label our emotions, activity in the amygdala decreases, and control shifts back to the prefrontal cortex, the part of the brain responsible for reasoning, decision-making, and emotional regulation [8]. In other words, naming the emotion helps us reclaim our mental clarity.

This process doesn't eliminate the discomfort; it indeed creates space between the feeling and the reaction. It allows us to observe rather than be consumed. And in that space, we begin to regain control, not by suppressing emotion, but by understanding it.

Strategies

When a shock hits, the emotional aftermath can feel overwhelming. But there are gentle, intentional practices that help us regain a sense of control and stability. These strategies are not about fixing everything instantly; they're about creating space to breathe, reflect, and begin the process of healing.

- Name the emotion
  Start by identifying what you're feeling. Say it aloud or write it down in a journal. For example: "I feel anxious right now", or "I'm overwhelmed and afraid". This simple act of naming the emotion, known in neuroscience as *affect labelling*, helps reduce its intensity. It signals to the brain that you are observing the emotion rather than being consumed by it. This shift activates the prefrontal cortex, allowing for clearer thinking and emotional regulation.
- Use breathing techniques
  One of the most effective ways to calm the nervous system is through intentional breathing. Try box breathing: inhale for four seconds, hold for four seconds, exhale for four seconds, and hold again for four seconds. Repeat this cycle several times. This technique engages the parasympathetic nervous system, which helps slow the heart rate, reduce stress hormones, and bring the body back to a state of calm.
- Practice grounding rituals
  Grounding helps anchor you in the present moment, especially when your thoughts are racing or your emotions feel chaotic. You might touch a familiar object, like a stone or a piece of fabric with texture. You can also look around and name five things you see, or listen to calming music that brings you comfort. These sensory cues remind your body and mind that you are safe, here, and now.

- Take micro-breaks
  Step outside, even if just for two minutes. Feel the air on your skin, notice the sounds around you, or let your eyes rest on something natural: a tree, the sky, a patch of grass. These small resets interrupt the emotional spiral and give your nervous system a chance to recalibrate. They don't solve the problem, but they help you return to it with more clarity and less tension.

REFLECTION

After redundancy struck, it felt as though the ground had vanished beneath steady feet, a sudden freefall into uncertainty. Thoughts spiralled like a storm, tightening the chest, making every breath feel borrowed. In that chaos, a small instinct surfaced: step outside. Bare feet pressed into cool grass, blades bending gently underweight. It was such a tiny act, almost laughably insignificant against the enormity of loss. And yet, the sensation of earth beneath skin whispered something profound: You are still here. For two quiet minutes, the future stopped screaming. The shock stopped replaying. There was no résumé to update, no frantic plan to hatch, just a human being, standing barefoot on solid ground, breathing. That moment didn't erase the uncertainty or mend the broken plans. But it mattered. It carved out a foothold in the avalanche, a fragile yet undeniable reminder that even in upheaval, presence is possible. And from presence, beginnings are born.

Mental/Cognitive Absorption

When a sudden shock occurs, whether it's a job loss, a breakup, or an unexpected crisis, our thoughts often become scattered and chaotic. The mind struggles to find order. It loops endlessly through "what if" scenarios, attempts to multitask without success, and finds it nearly impossible to focus on even the simplest task. This mental fog is not just frustrating; it's a natural response to cognitive overload.

Scientific Insight

Psychologists refer to this state as cognitive load overflow [9]. Our working memory, which is responsible for holding and processing information in the short term, becomes overwhelmed. When too many thoughts compete for attention, decision-making quality drops significantly. Neuroscience research shows that multitasking under stress can increase error rates by as much as 50% [10]. The brain, in survival mode, prioritises perceived threats over logical reasoning, making it harder to think clearly or act effectively.

STRATEGIES FOR MENTAL RECOVERY AND CLARITY

To regain mental clarity and reduce overwhelm, we can turn to a few simple but powerful strategies. These practices help restore focus, calm the nervous system, and create space for thoughtful action.

- Single-tasking with intention
  Instead of trying to do everything at once, choose one small, manageable task and commit to it fully. It could be something as simple as making a cup of tea, replying to one email, or tidying a small corner of your space. Completing even a minor task successfully can anchor you in the present and remind you that you are still capable of action.
- Externalise your thoughts
  When your mind feels cluttered, take a moment to write down your worries, questions, or racing thoughts in a notebook or journal. This act of externalising helps clear your working memory and gives you a visual map of what's occupying your mental space. It's not actually solving everything, but it creates distance between you and the chaos.
- Time-box your focus
  Set a timer for a short, defined period, say, 15 minutes, and choose one task to focus on during that time. When the timer ends, stop. This technique prevents spiralling and helps you stay grounded in manageable chunks of time. It's especially helpful when motivation is low or when everything feels too big to tackle.
- Reframe your thinking
  Ask yourself a grounding question: *What do I know right now, and what am I imagining?* This helps separate facts from fears. Often, our minds fill in gaps with worst-case scenarios. By identifying what's real versus what's speculative, we reduce anxiety and regain perspective.

REFLECTION

On the day of the layoff, thoughts scattered like shards of glass, sharp, chaotic, impossible to gather. Concentration dissolved into fog. Decisions felt unreachable, as if locked behind a wall of noise. In that haze, one simple intention emerged: make a shopping list and buy groceries. Nothing grand, nothing life-altering, just one tangible act. Walking through the aisles, choosing apples, ticking items off the list, something subtle shifted. For a brief stretch of time, the storm quieted. There was no spiralling about the future, no replaying the shock, only the ordinary rhythm of a human being moving through a store, selecting food, and breathing. That moment didn't erase uncertainty or restore what was lost. But it mattered. It whispered a truth easily forgotten in crisis: not everything is broken. Function still exists. Capability still lives. Presence remains. And from that presence, the first thread of stability begins to weave its way back.

Physical Absorption

The body remembers shock. It doesn't just pass through us, it stays. Muscles tighten as if bracing for impact: shoulders rise towards the ears, the jaw locks like a vice, the stomach twists and churns. These are not random reactions; they are the body's ancient survival mechanisms at work.

The science behind it:

> According to the polyvagal theory, the vagus nerve plays a central role in regulating our stress response [11]. When we experience a threat, physical or emotional, the sympathetic nervous system surges into action, triggering the classic fight-or-flight state. Heart rate spikes, digestion slows, and energy floods the muscles. This is adaptive in the short term, but when stress persists, the body struggles to return to balance. That's where the parasympathetic branch, the "rest-and-digest" system, comes in. Activating it is key to calming the storm and restoring equilibrium.

Physical actions are distractions; they are direct interventions. They send powerful signals to the nervous system that it is safe to downshift from high alert.

Practical strategies to reset your system:

- Progressive muscle relaxation: methodically tense and release each muscle group, from your toes to your forehead. This helps discharge stored tension and reminds the body what relaxation feels like.

- Movement: gentle activity, walking, stretching, light exercise – burns off excess adrenaline and restores a sense of groundedness.
- Hydration and nutrition: shock depletes glucose and fluids. Replenish with water and protein-rich foods to stabilise energy and mood.
- Sleep hygiene: even if deep sleep feels out of reach, prioritise rest. Consistent routines, dim lights, cool room, no screens, signal safety to the nervous system and promote recovery.

A NOTE

After the breakup, grief didn't just live in the heart, it colonised the body. Every muscle seemed to carry its weight. The jaw locked so tightly it ached, as if holding back words that would never be spoken. Headaches became constant companions, pulsing behind the eyes like a silent alarm, reminding that sorrow isn't confined to thoughts, it burrows into bones, settles deep in tissue.

One night, exhausted from the tension, a decision was made to try something different. It began with the smallest gesture: gentle jaw stretches, fingertips tracing the tightness that had been ignored for weeks. Then came progressive muscle relaxation, slow, deliberate, moving through each part of the body, teaching it the forgotten language of release. At first, it felt awkward, almost futile, like coaxing a locked door to open with whispers. But night after night, the practice continued.

And then, something shifted. Within a week, headaches softened. The vise around the jaw loosened. It was as if the body exhaled for the first time in months.

That experience revealed something profound: healing isn't only emotional, it's physical. The body holds stories, heartbreaks, and fears like secret archives. And sometimes, the first step towards feeling whole again is listening, not to the mind's noise, but to what the body has been trying to say all along.

Digital Absorption

In today's hyperconnected world, shocks don't just come from real-life events, they're amplified by screens. A single notification can feel like a bounce to the nervous system. A sudden text message, a breaking news alert, or a viral comment on social media can send ripples of anxiety through the body. What used to be occasional disruptions have become constant micro-interruptions, each one demanding attention, and each one pulling us out of balance.

The science behind it:

Research shows that the relentless stream of notifications elevates cortisol, the body's primary stress hormone, while fragmenting our focus [12]. These digital "micro-shocks" may seem small, but they accumulate silently, leaving us more fragile and less equipped to handle bigger emotional blows. Over time, this constant state of alert weakens our capacity to absorb and process real-life shocks.

Practical strategies to reclaim calm:

- Digital pause: Silence notifications during periods of recovery. Give your nervous system space to breathe.
- One trusted source: When global crises or breaking news hit, resist the urge to doomscroll. Choose one reliable outlet and check it at set times.
- Delay responses: Before reacting to a provocative message or email, wait 24 hours. Time creates perspective.
- Create a digital sanctuary: For short intervals, keep your phone in another room. Let your mind rediscover stillness.

A REFLECTION

After receiving a harsh email during a workplace conflict, adrenaline surged like a fire alarm. Fingers hovered over the keyboard, ready to unleash a response sharp enough to cut through the sting. Every word felt justified in that moment, anger demanding an outlet. But then, a pause. The laptop closed with a quiet click, and a 12-hour delay was set. Sleep became the buffer between impulse and intention. The next morning, the draft was revisited, and it felt reckless, almost foreign, like reading the thoughts of someone else. That pause had done its work. It saved the situation from spiralling into escalation. That experience revealed something profound: absorption isn't passive. It's an active, deliberate choice, a shield for peace. It's the discipline to create space between trigger and response, because in that space, clarity lives.

Social Absorption

Social interactions during times of shock can be a lifeline, or a landmine. When life blindsides us, the people around us can either help us regain our footing or push us further off balance. A single conversation can soothe the nervous system, or it can send stress levels soaring.

The science behind it: Research consistently shows that genuine social support lowers cortisol, the body's primary stress hormone, and even accelerates cardiovascular recovery after stressful events. In other words, the right kind of connection isn't just comforting; it produces measurable physiological benefits. Conversely, toxic interactions, such as gossip, criticism, or judgement, can intensify stress, leaving us more depleted and less resilient. Later, we'll explore toxic dynamics in depth: how to recognise them, understand their impact, and take deliberate steps to remove ourselves from those situations.

Practical strategies for navigating social dynamics during shock:

- Reach out selectively: Choose people who are supportive, calm, and nonjudgemental. Not everyone deserves access to your vulnerability.
- Set boundaries: Limit exposure to gossip, blame, or negativity. Protect your mental space like you would a physical wound.
- Use "containment conversations": Call a trusted friend and say, *I don't need advice, I just need to vent for five minutes.* This sets expectations and prevents overwhelm.
- Reciprocal support: Sometimes, helping someone else through their shock can buffer your own stress. Shared humanity is grounding.

A REFLECTION

When the contract fell through, it felt like the floor had vanished beneath steady feet. Panic surged like a tidal wave, crashing against the chest, followed by the heavy, suffocating weight of self-doubt pressing down like a storm cloud. The instinct was immediate and primal: retreat, curl inward, shut the world out. Isolation whispered promises of safety, at least, that's what the mind insisted. But beneath the noise, a quieter voice stirred: reach out instead. With trembling hands, the phone was lifted, a number dialled almost on autopilot. When the call connected, words refused to form. All that escaped was a fragile sentence: "I just need you to listen". There was a pause, a soft, grounding silence, and then three simple words: "I'm here, always". No advice. No judgement. No attempt to fix what couldn't be fixed. Just presence. Just care. And that was enough. In that stillness, something inside loosened. The panic didn't vanish, but it softened, like a storm losing its edge. That moment revealed a truth often overlooked: healing begins not with grand gestures or perfect words, but in the quiet company of someone who truly sees you. Sometimes, the most powerful thing we can hear is simply: You're not alone.

Professional Absorption

In the modern workplace, shocks are almost inevitable. They arrive without warning: an unexpected audit, sudden layoffs, a safety incident on the floor, or the slow, corrosive effect of toxic leadership. Each one sends ripples through the system, unsettling routines and triggering stress responses. Absorbing these shocks isn't just about personal resilience; it requires a combination of individual strategies and structural support.

THE SCIENCE BEHIND IT

Occupational health psychology highlights the Job Demands-Resources (JD-R) theory [13], which explains that when job demands spike, such as during crises, having adequate resources acts as a buffer. These resources can be tangible, like flexible policies and clear communication, or relational, like supportive managers and cohesive teams. When these are in place, the impact of workplace shocks is significantly reduced.

Practical strategies for navigating workplace shocks:

- Pause before reacting: Resist the urge to fire off an email or make a hasty decision in the heat of panic. A few minutes of stillness can prevent hours of damage control later.
- Clarify the facts: In moments of chaos, assumptions multiply. Ask, *What exactly is required of me right now?* This question cuts through the noise and restores focus.
- Take micro-breaks: Step outside for fresh air, grab a glass of water, or simply stretch. These small resets help regulate your nervous system and prevent overwhelm.
- Seek allies: Absorbing stress is easier when shared. Lean on your team, delegate where possible, and create a sense of collective resilience.

A REFLECTION

When the unexpected safety audit hit, tension spiked like electricity in the air. The factory floor felt heavy, thick with anxiety, with every glance, every movement charged with urgency. Instinct screamed for frantic action: scramble, fix, control. But instead, something different happened. The team was gathered for a three-minute breathing exercise. It felt awkward, almost absurd – breathing exercises in a factory? Yet, as the first inhale stretched deep and the first exhale released, the atmosphere began to shift. Shoulders dropped. Jaws unclenched. The rigid edges of panic softened. That small pause didn't just calm nerves, it restored composure. It gave the team what they truly needed: clarity, steadiness, and the confidence to face the inspectors without fear. That moment taught a powerful truth: sometimes the most effective leadership move isn't doing more, but creating space to breathe.

INTEGRATING SCIENCE AND STORY

In the modern workplace, shocks are not rare; they are part of the landscape. They arrive without warning, like sudden storms on what seemed like a clear day: an unexpected audit, a round of layoffs announced in a terse email, a safety incident that halts production, or the slow, corrosive drip of toxic leadership that erodes trust and morale over time. Each of these moments sends ripples through the system, unsettling routines, shaking confidence, and triggering stress responses that can feel overwhelming.

When these shocks hit, they don't just disrupt schedules; they disrupt people. They unsettle the nervous system, cloud judgement, and make even simple decisions feel monumental. Absorbing these shocks isn't about pretending to be unshakable or "toughing it out". It's about creating space to respond instead of reacting. And that requires more than personal grit; it calls for a combination of individual strategies and structural support.

THE SCIENCE BEHIND IT

Occupational health psychology offers a powerful lens through the JD-R theory [14]. This framework explains that when job demands spike, such as during a crisis, our ability to cope depends on the resources available to us. These resources can be tangible, such as flexible policies, clear communication, and fair workloads, or they can be relational, like supportive managers, psychologically safe teams, and a culture that values well-being over constant urgency.

When these resources are present, they act as shock absorbers, buffering the impact of stress and preventing burnout. When they're absent, even small disruptions can feel catastrophic. This is why resilience isn't just an individual trait; it's a shared responsibility between people and the systems they work within.

Practical Strategies for Navigating Workplace Shocks:

- Pause before reacting: In the heat of panic, the impulse to act fast can be overwhelming. But speed often breeds mistakes. Take a breath. Step back. A few minutes of stillness can prevent hours, or even days, of damage control later.
- Clarify the facts: Chaos breeds assumptions, and assumptions breed anxiety. Ask the grounding question: *What exactly is required of me right now?* This simple inquiry cuts through the noise and restores focus.
- Take micro-breaks: When stress spikes, the body needs signals of safety. Step outside for fresh air. Grab a glass of water. Stretch your shoulders. These small resets regulate your nervous system and keep you from spiralling.
- Seek allies: Absorbing stress is easier when shared. Lean on your team. Delegate where possible. Create a sense of collective resilience. When we carry the weight together, it feels lighter.

A REFLECTION

The day the compliance review was announced, it felt like the air in the office thickened instantly. Conversations clipped short. Faces tightened. You could almost hear the cortisol humming beneath the surface. Instinct screamed: move faster, fix everything, and control the chaos. But then, pause. Instead of feeding the frenzy, the team was gathered for something unexpected: a three-minute breathing exercise. At first, it felt awkward, almost absurd. Breathing exercises? In a factory? Yet, as the first inhale stretched deep and the first exhale released, something shifted. The tension began to melt. Shoulders dropped. Jaws unclenched. For the first time that day, space opened for clear thought. That small pause didn't just calm nerves; it changed the trajectory. It gave the team composure, steadiness, and the confidence to face the challenge without fear. It was a reminder that resilience isn't about being untouched by stress; it's about creating islands of calm in the storm. Sometimes, those islands are as simple and as radical as breathing together.

CLOSING NOTES

The absorption period is not a mere pause; it is a profound act of resilience. It is the art of holding steady when the ground beneath you feels like it's crumbling. It is the sacred, fragile space between the impact of a shock and the first steps towards recovery. In this space, we do something radical: we allow ourselves to stop running. We give ourselves permission to breathe, to feel, to acknowledge the weight of what has happened without letting it consume us.

Absorption is not about denial. It is not about pretending everything is fine. It is about saying, *This hurts. This is hard. And yet, I am still here.* It is the quiet strength that whispers when the world is shouting: *Hold on. This wave will pass. You are not broken. You are still standing.*

When performance falters, when plans unravel, when the unexpected crashes into your life like a storm, absorption becomes your anchor. It is not glamorous. It does not make headlines. But it is the difference between being swept away by chaos and finding a foothold in the midst of it. Absorption is the deep breath before the next climb, the stillness that makes movement possible again.

And here is the truth: without absorption, there can be no recovery. Without this pause, there is no bridge to resilience. We cannot leap from shock to growth without first honouring the space in between, the space where we gather our strength, where we remember who we are, where we choose to keep going.

In the next chapter, we will explore adaptation and recovery, the stage where stability transforms into growth, where the rawness of shock begins to shape resilience. But never forget this: the journey to strength begins here, in the quiet, courageous act of holding steady.

QUICK PRACTICE: NAVIGATING THE ABSORPTION PERIOD

Step 1: Recall a Shock

Think of a recent moment when you were blindsided: a tough email, a conflict, or unexpected bad news.

My Shock: ______________________________
(Example: My manager suddenly criticized my work in a meeting.)

Step 2: Notice Your First Response

Reflect on how your body and mind reacted in the first minutes.

- What emotions surged?
- What did you feel in your body?
- Did you act immediately, freeze, or withdraw?

Step 3: Contain, Don't Collapse

Write down one thing you did, or could have done, to hold steady without escalating the situation.

Containment Strategy: ______________________________
(Example: I paused, breathed slowly, and reminded myself not to reply instantly.)

Step 4: Identify Your Buffer

List the resources or supports that helped you absorb the shock.

- Who or what steadied you? (a friend, walking, journaling, breathing, etc.)

Step 5: Small Adjustment for Next Time

Choose one micro-change you'll try in your next disruption.

My Commitment: ______________________________
(Example: Next time, I'll step outside for two minutes before responding.)

Tip: This practice is not about solving everything in the moment. It's about buying space. Each time you pause, contain, and reflect, you train your system to hold steady in the aftershock; the very heart of resilience.

REFERENCES

1. Orji, L.C. and Ita, U.A., 2024. Amygdala hijack: Contemporary insights into causes, correlates and consequences. *African Journal of Health, Nursing and Midwifery*, 7(3), pp.102–111.
2. Goleman, D. and Boyatzis, R., 2008. Social intelligence and the biology of leadership. *Harvard Business Review*, 86(9), pp.74–81.
3. Wang, X., Ning, R. and Zhao, X., 2023. Generalized mixed shock model for multi-component systems in the shock environment with a change point. *Proceedings of the Institution of Mechanical Engineers, Part O: Journal of Risk and Reliability*, 237(4), pp.619–635.
4. Sebat, F., Musthafa, A.A., Johnson, D., Kramer, A.A., Shoffner, D., Eliason, M., Henry, K. and Spurlock, B., 2007. Effect of a rapid response system for patients in shock on time to treatment and mortality during 5 years. *Critical Care Medicine*, 35(11), pp.2568–2575.
5. Clear, J., 2018. *Atomic habits: Tiny changes, remarkable results*. Penguin Publishing Group.
6. Housel, M., 2020. *The psychology of money: Timeless lessons on wealth, greed, and happiness*. Harriman House Limited.
7. Burklund, L.J., Davies, C.D., Niles, A., Torre, J.B., Brown, L., Vinograd, M., Lieberman, M.D. and Craske, M.G., 2024. Affect labeling: A promising new neuroscience-based approach to treating combat-related PTSD in veterans. *Frontiers in Psychology*, 15, p.1270424.
8. Bonnet, L., Comte, A., Tatu, L., Millot, J.L., Moulin, T. and Medeiros de Bustos, E., 2015. The role of the amygdala in the perception of positive emotions: An "intensity detector". *Frontiers in Behavioral Neuroscience*, 9, p.178.
9. Skulmowski, A. and Xu, K.M., 2022. Understanding cognitive load in digital and online learning: A new perspective on extraneous cognitive load. *Educational Psychology Review*, 34(1), pp.171–196.
10. Zehra, S.R. and Malik, M., 2025. The cognitive cost of multitasking in high-stress professions: Implications for mental efficiency, error rates, and long-term cognitive health. *The Critical Review of Social Sciences Studies*, 3(2), pp.2469–2487.
11. Porges, S.W., 2022. Polyvagal theory: A science of safety. *Frontiers in Integrative Neuroscience*, 16, p.871227.
12. Venter, F.P., 2019. *The effect of receiving mobile text messages on salivary cortisol levels in physiology students at the University of the Free State* (Doctoral dissertation, University of the Free State).
13. Bakker, A.B. and Demerouti, E., 2014. Job demands-resources theory. In *Wellbeing: A complete reference guide*. John Wiley, pp.1–28.
14. Bakker, A.B., Demerouti, E. and Sanz-Vergel, A., 2023. Job demands-resources theory: Ten years later. *Annual Review of Organizational Psychology and Organizational Behavior*, 10(1), pp.25–53.

4 The Climb Back to the Ridge

Adaptation and Recovery in the Vulnerability Period

Where there is ruin, there is hope for a treasure.

Rumi

A PRELUDE

A morning arrived when everything that had been built over years seemed to vanish behind a single closed door. The office that had been entered day after day, the systems that had been carefully shaped, the safety procedures carrying a familiar signature, all of it was pushed to a distance in the space of one brief, rehearsed meeting. The language was polished, the tone controlled, the decision presented as inevitable. A position was "restructured", "realigned", gently moved to the margins and then beyond them. The company, proudly framed as a model of efficiency and profit, no longer recognised any need for the human being behind the metrics.

A cardboard box was placed in tired hands, filled with fragments of a working life: a mug, a notebook, a few photos, and a hi-vis vest that had seen too many long shifts. The building's glass façade caught the morning light, the logo gleaming as if nothing in the world had shifted. Yet inside, a storm was already breaking. Years of devotion to safety work,

DOI: 10.1201/9781003789895-5

time in fire zones, nights spent drafting emergency plans, days spent training teams to act when seconds mattered, were reduced to a polite farewell and a scripted email. The machinery that had once opened its doors now stood cold and distant, continuing to hum along without so much as a pause.

Still, collapse did not come. Surrender did not take hold. In the space left by shock and humiliation, something older than fear and stronger than shame began to rise. It was the quiet pride of someone who had stood for more than numbers on a dashboard. The work had never been only about targets or compliance; it had been about dignity, about the belief that systems exist to protect people, not discard them when budgets tighten or strategies change. That belief did not vanish with the swipe of an access card. It remained, stubborn and steady, beneath the hurt.

The months that followed unfolded as a long uncertainty period. The structure that once organised every day – rosters, meetings, reports, audits – dissolved almost overnight. Mornings became strangely open, and evenings stretched out, edged with worry. A professional identity once taken for granted had to be redefined from the ground up. Questions appeared that had never needed to be asked before: Who is the safety professional without the company logo? What remains when the title is removed from the email signature?

Within that disorientation, something vital slowly surfaced. Resilience began to show itself, not as a simple return to what used to be, but as the gradual creation of something that could hold even when the ground shook again. New routines were tested. Skills were re-examined and repurposed. Different doors were knocked on. Small opportunities were accepted, then expanded. The fall, painful, became a starting point rather than an ending. Adaptation was not sudden; it came through ordinary days, through doubt, through quiet persistence. Yet, step by step, a new path began to form. The climb had started.

This chapter is drawn from that climb. It is shaped by the vulnerability period, that stretch of time when the fall has ended but the footing is still unstable, when pride has been bruised but not erased. The discussion is informed by theory and models, yes, but it is also grounded in lived experience: the feeling of standing in the aftermath of loss and consciously choosing not to stay there. What follows is a reflection on adaptation and recovery, on the strength that can emerge when deeply held values meet disruption, and on the quiet power of rebuilding with intention, one small decision at a time.

****

The storm has passed, and the freefall has stopped, but the earth beneath your feet still trembles. In the wake of a major disruption, there is a vulnerability period, a stretch of time when you are no longer plunging but not yet surefooted. Your PF whether measured as physical output, emotional stability, cognitive clarity, financial security, or professional productivity, sits below your baseline. The shock has been absorbed, as discussed in Chapter 3, but the gap between where you are and where you need to be remains wide. You cannot simply wait for things to feel normal. Recovery is not a passive drift back up the slope; it is an active, effortful climb. The path across the gap involves two distinct but overlapping processes: adaptation and recovery. This chapter explores that climb, showing how you pivot under load and then build back up, drawing on recent research, engineering theory, psychology, disaster management, and personal experience.

Resilience researchers note that resilience is the capacity to resist or rapidly recover from impacts, and it entails learning and adapting to reduce future vulnerabilities [1]. Engineering definitions have evolved accordingly. The North American Transmission Forum (NATF) updated its resilience definition for electric systems to include five elements: prepare, anticipate, absorb, adapt, and recover, underscoring that adaptation and recovery are distinct activities. Psychological resilience theories likewise distinguish between the ability to recover and the positive adaptation to life conditions [2]. In other words, after a shock there is first a period where we must modify ourselves or our systems to function under new conditions (adaptation), and then a period where we climb back towards our original capacity (recovery). These two processes happen as we are still vulnerable to secondary shocks. Understanding how they work is the key to bridging the gap.

The career behind this book has unfolded at the intersection of theory and practice. Trained as a safety engineer and risk consultant, years have been spent helping organisations prepare for and respond to crises in high-risk settings such as oil and gas, power generation, and firefighting. Alongside this professional work, personal disruptions have been endured: unexpected job loss, professional demotion, physical injuries, each one forcing resilience principles to be tested in real life rather than left on paper. In the chapters which follow, research findings are placed alongside lived experience, allowing the vulnerability period to be examined through multiple lenses. The aim is to offer grounded guidance for readers navigating their own post-shock terrain, where identity feels unsettled, structures have shifted, and a new form of stability is still in the making.

UNDERSTANDING THE VULNERABILITY PERIOD

The Resilience Lifecycle: Beyond Response and Recovery

Traditional disaster management frameworks follow a linear sequence: mitigation → preparedness → response → recovery. The underlying assumption is that after the crisis is contained, the system simply returns to normal. But researchers and practitioners increasingly recognise that this model is too simplistic. The NATF/Electric Power Research Institute (EPRI) companion document explains that resilience activities begin long before the crisis and includes Ref. [3]. After the initial shock has been absorbed, recovery does not start with the flick of a switch; instead, systems enter a vulnerability period where adaptation and recovery overlap. Adaptation is the pivot that allows the system to function, albeit at reduced capacity, under changed conditions. Recovery is the climb that gradually restores capacity and, ideally, reduces vulnerability to future shocks. In engineering terms, the resilience curve resembles a trapezoid: a sharp drop after the event (shock), a flat stretch as the system absorbs the impact, an upward slope as adaptation takes hold, and a longer slope toward baseline as recovery proceeds.

The COVID-19 pandemic offered a real-world illustration of the vulnerability period. Across the world, societies oscillated between emergency response and attempts to reopen, revealing a missing stage in the resilience cycle. A structural-engineering article on community resilience argued that disaster management models should include an adaptation phase between response and recovery to reflect "prolonged duress", a period when communities must slow ongoing damage by adjusting their recovery strategies. The authors proposed replacing the term "recovery" with "adapt" in certain contexts because a simple return to the pre-disaster state may re-embed vulnerabilities; adaptation encourages innovation and new targets for functionality. Inserting adaptation into the cycle, we acknowledge that a period of flexible adjustment is necessary before full recovery can begin.

Psychological models emphasise that resilience is a dynamic process, rather than an outcome. A comprehensive literature review on psychological resilience notes that definitions vary, but many share a core idea: resilience involves recovering from disruption, adapting to change, and bouncing back with new insight. The review lists five components: the ability to recover, enduring adaptive traits, capacity to bounce back, dynamic process over time, and positive adaptation to life conditions. Thus, adaptation and recovery are integral parts of a continuous process. A recent theoretical model, the Wither or Thrive Model of Resilience [5], integrates multiple resilience frameworks and emphasises the distinction between a trajectory of resilience and a trajectory of recovery. The model acknowledges that adaptive and maladaptive functioning exist on a continuum and that individual, contextual, and environmental factors interact to influence whether someone follows a resilient trajectory (maintaining function or even thriving) or a recovery trajectory (initial decline followed by rebound). This underscores that adaptation and recovery are separate but related processes within the vulnerability period.

During the long work of recovery, resilience began to appear less as a fixed personal trait and more as something that grows through connection and deliberate action. This view is captured in what is called the Resilience Activation Framework. In that model, resilience is described as the capacity of a person, a team, or a system to endure and recover from events that threaten stability, direction, or growth. The idea is not abstract. Many lives, many careers, carry moments when the ability to function is pushed to its edge, when sleep is broken, concentration frays, and even simple decisions feel heavy. In those periods, progress is rarely sustained by inner strength alone. Supportive relationships, practical structures, and small intentional steps each day all play a part in helping recovery take shape and in turning shock into a starting point rather than an endpoint.

The framework outlines two essential elements: the disruption itself and the system's ability to respond. That response emerges through patterns of change that allow continued movement, even when the path is uncertain. These patterns rely on different forms of capital: human, emotional, financial, social, and political. Each one plays a role in shaping the way we adapt and rebuild.

During the vulnerability period, it became clear that access to support could tilt the entire recovery path. A mentor's quiet insight, a moment of clarity journaling, or a brief gesture of encouragement from a friend, these were more than simple comforts. They acted as catalysts, small turning points that helped restore focus, enabled decisions, and marked the first steps of the climb back from shock. Within the Resilience Activation Framework, resilience is understood as something that can be activated through intentional support. That perspective reshapes recovery: progress is no longer left to time and willpower alone, but is linked to consciously reaching for the resources, relationships, and practices that help movement resume with purpose.

The Window of Vulnerability

The vulnerability period begins once the shock has been contained and the rapid decline in PF has been halted. In this period, the system is no longer crashing but remains far from its baseline. Buffers are depleted, reserves are low, and secondary shocks, even minor ones, can have outsized impacts. This window is uncomfortable, and represents the opportunity space for adaptation and recovery. If adaptation is quick and strategic, the window shrinks, and recovery can proceed smoothly. If adaptation is haphazard, the window stretches, leaving the system exposed and prolonging suffering.

One of the most disorienting vulnerability periods began in the middle of a project that had absorbed almost a year of focused effort. A safety redesign was underway for a regional energy facility, a role demanding precision, coordination, and constant vigilance. The work was intricate, the stakes were high, and strong momentum had been built. Then, without warning, the project was shelved after a shift in corporate priorities. Funding disappeared. The team was reassigned. The role at the centre of the work was quietly dissolved.

The impact was immediate. A deep sense of displacement set in. The calendar that had been full of meetings, site visits, and deadlines emptied almost overnight. The daily rhythm that had given structure and meaning to each week simply vanished. A glowing computer screen remained, but the tasks it once represented no longer existed. The sense of purpose carried into each toolbox talk, each redesign meeting, and each incident review had evaporated in a matter of days. Colleagues had to be informed, schedules reorganised, and families told that the work which had seemed so important had suddenly been halted.

In the days that followed, PF dropped sharply. Energy levels fell. Focus scattered. Confidence wavered. Direction became blurred. The project had been more than a piece of work; it had served as an anchor for professional identity. Without it, even small pressures gained weight. A delayed payment, a minor health concern, and an offhand remark from a peer could now unsettle the entire day. The vulnerability period had fully begun.

Adaptation was no longer optional; it became essential. A new routine was drafted, mentors were contacted, and adjacent fields and roles were quietly explored. The experience was gradually reframed, from a simple loss into a transition phase. Recovery did not arrive in a single breakthrough. It emerged through small, almost ordinary steps: learning, reconnecting, rebuilding, and testing new directions. That period exposed how fragile stability can be, and how powerful it is to reclaim it through deliberate, steady action.

ADAPTATION: THE PIVOT UNDER LOAD

What Adaptation Means

Adaptation refers to modifying behaviours, structures, and strategies so that a system can operate under changed conditions. Unlike recovery, which aims to restore lost capacity, adaptation is about creating a new way of functioning within constraints. This may involve developing new routines, reframing goals, reallocating resources, and redefining roles. Research across disciplines emphasises adaptation as a core component of resilience. Engineering guidelines describe adaptation as the flexibility and scalability of systems during an event. In psychology, adaptation is described as positive adaptation to life conditions, the ability to engage with life, and make meaning even when facing chronic stress or trauma. The Wither or Thrive model specifically distinguishes between resilient trajectories (where adaptation maintains function) and recovery trajectories (where adaptation follows an initial decline). Thus, adaptation is a process and outcome, enabling continuation of function in the face of altered realities.

Adaptation is not one thing; it is a set of related mechanisms. It may be cognitive, emotional, physical, social, and organisational. It can be deliberate (planned changes) or emergent (improvisations that evolve organically). It may involve transformation (a fundamental shift in purpose or identity) or adjustment (a tweak to how something is done). Crucially, adaptation is often messy. It emerges from trial and error, from sensing what the new landscape demands, and experimenting with solutions. Adaptation is time-bound. In the vulnerability period, adaptation must happen quickly enough to prevent further decline, but thoughtfully enough to avoid entrenching poor habits. As the Resilience Activation Framework argues, negative outcomes after disasters often result from a time-limited stress response followed by resilient adaptations. In other words, after the immediate shock, there is a narrow window where adaptation can either set the stage for recovery or open the door to further decay.

Why Adaptation is Essential

Adaptation matters for several reasons:

1. *You cannot recover without adjusting.* Simply replicating what you did before the disruption will not work because the world has changed. In the aftermath of a disaster, attempting to rebuild exactly what was lost can re-embed vulnerabilities. Whether you are an individual, an organisation, or a community, you must respond to new constraints and opportunities.
2. *Adaptation shortens the vulnerability window.* By pivoting quickly, you reduce the time you spend in a fragile state. Quick adaptation speeds the onset of recovery and lowers the risk of secondary shocks. In the NATF/EPRI framework, adaptation is emphasised as a means-objective which allows systems to deliver capability under changing conditions.

3 *Adaptation reduces future vulnerability.* By learning from the disruption and making strategic changes, you emerge better prepared for the next shock. Rapid return to the status quo can be dangerous because it fails to address systemic weaknesses. Adaptive modifications like decentralising supply chains, diversifying revenue streams, or cross-training staff make you more robust.
4 *Adaptation preserves identity and meaning.* Humans draw a deep sense of identity from roles and routines. When those structures are disrupted, adaptation becomes the bridge which holds a coherent life story together. A sudden job loss, for example, can be transformed through reframing: "full-time employee" becomes "consultant and learner", not as a comforting slogan but as a new, workable identity. In this way, purpose is not erased by disruption; it is redirected, allowing meaning to survive even when circumstances shift.
5 *Adaptation mobilises resources and social capital.* The Resilience Activation Framework posits that access to human, social, economic, and political capital can activate resilience processes that promote positive adaptation. In other words, adaptation taps into networks of support and draws upon collective strength.

Mechanisms of Adaptation

Adaptation can be understood through a set of mechanisms which operate at both individual and organisational levels. Across practice and lived experience, seven levers stand out as particularly powerful in shaping how recovery, and growth unfold:

1 *Sense-making and information management.* You cannot adapt effectively until you understand what has changed, what remains the same, and where the uncertainties lie. This involves gathering accurate data, consulting experts, and avoiding rumours. During the COVID-19 pandemic, leaders who adapted quickly were those who obtained clear, up-to-date information on the virus, supply chains, and government policies and acted on that knowledge. In my personal job loss, sense-making involved assessing my skills, reviewing my finances, and understanding the labour market. This clarity prevented me from making knee-jerk decisions.
2 *Prioritisation under constraint.* During the vulnerability period, capacity is limited and not everything can be carried forward. Choices must be made about which activities hold the highest value, and those must be protected, while lower-value tasks are deferred or

suspended. When a gas turbine failure occurred, the priority became restoring safe partial power output, and non-critical maintenance was deliberately postponed. In a parallel way, after a sudden job loss, attention narrowed to essential expenses, with discretionary items such as holidays and other luxuries set aside. In both cases, adaptation began with disciplined prioritisation: accepting that not all loads can be carried, and choosing carefully which ones must remain.

3 *Resource reallocation.* Adaptation often requires shifting money, people, time, and attention towards the new bottlenecks. In the compressor incident, the strongest mechanical engineers were reassigned to the failure analysis, budgets were pulled from less urgent projects, and external consultants were brought in. On a more personal level, the hours once spent commuting were redirected into job searches and skill development. In both cases, resources were deliberately re-routed to where the system was weakest, allowing recovery to begin at the actual point of constraint rather than at the edges.

4 *Minimum viable routine (MVR).* When everything feels chaotic, a simple, repeatable routine provides stability. For me, this meant waking at the same time each day, setting small daily goals (two job applications, one hour of exercise), spending an hour on professional reading, and connecting with one supportive person. At the plant, our MVR comprised daily briefings, hazard checks, and progress reviews. MVRs stabilise PF by creating structure when mental bandwidth is low.

5 *Creating guardrails.* Guardrails are rules that prevent reactive decisions that could worsen the situation. At the plant, we instituted a rule that no modifications to any turbine could be made without peer review, preventing impulsive fixes that might introduce new hazards. On a personal level, guardrails might include not replying to emotionally charged emails immediately, or capping discretionary spending to avoid financial crises.

6 *Redundancy and backup.* Adaptive systems deliberately build redundancy – alternative suppliers, backup communication channels, and secondary transport routes – so that core functions can continue when a primary pathway fails. After the incident, backup generators and emergency inventories were identified and strengthened to support continued operations during future disruptions. On a personal level, the professional network was broadened to include multiple potential employers and income streams, reducing dependence on any single role, and creating alternative pathways for work, income, and growth.

7 *Narrative reframing.* Interpretation of disruption strongly shapes the actions that follow. Resilient individuals and organisations deliberately reframe the story from "we have failed" to "we are learning" or "we are being tested". In the first adaptation briefing after the turbine failure, the message to the team was clear: this was not a catastrophe, but a stress test, and the task was to learn from it and build a more reliable system. Such reframing does not deny the severity of the event; instead, it turns attention towards agency and possibility. When disruption is viewed as a proving ground rather than an ending, creativity is unlocked, options multiply, and the path to recovery becomes easier to see.

Adaptation Across Domains

Emotional and Mental Adaptation

Emotionally, adaptation asks for two difficult tasks at once: recognising intense feelings and making meaning of the disruption. Positive psychology research points out that resilience is the capacity to "bounce back", and to adapt positively, to integrate the experience and move forward with a changed, and often deeper, understanding of self and context. In practice, this work can involve naming emotions, practising mindfulness, journalling, using cognitive reframing, or seeking therapeutic support. In one period of job loss, for example, there was a steady oscillation between hope and despair; progress began when fear and shame were named ("there is uncertainty here, and there is embarrassment"), when the event was reframed as a chance to realign work with core values, and when counselling was used to process the more complex layers of grief and anger. At the plant, emotional adaptation took the form of honest daily debriefs, where collective fear and fatigue were acknowledged openly, and the incident was framed as a learning opportunity rather than a professional stain.

Mentally, adaptation requires an update of the internal maps that guide action. Old assumptions must be loosened, sometimes abandoned, so that cognitive energy can be redirected from defending the past towards solving the problems now in front of the system. A previously assumed career trajectory, for instance, had to be released and replaced with the idea that new possibilities might lie outside familiar roles, and titles. In the technical domain, engineers were required to let go of long-held beliefs about the near-infallibility of certain safety mechanisms, and instead design new procedures which treated those systems as fallible and in need of backup. Emotional and mental adaptation, taken together, allow disruption to be absorbed, understood, and used as raw material for a more honest and robust way of operating.

Physical Adaptation

Physical adaptation focuses on how bodies and work environments are adjusted so that strain, injury, and fatigue do not quietly erode resilience. After a disruption, the same workload that once felt manageable can become overwhelming; systems that were "good enough" when everything was stable may no longer be safe. Physical adaptation therefore involves recalibrating effort, rest, posture, and exposure. This can mean restructuring shifts, redesigning tasks to reduce manual handling, introducing ergonomic tools, or embedding micro-breaks and stretching into daily routines so that the body is treated as critical infrastructure rather than an expendable resource. At the individual level, physical adaptation often includes physiotherapy, medical review, tailored exercise, sleep hygiene, and deliberate recovery periods after high-stress work. In one case, a professional used an enforced career break to address chronic lower back pain through physiotherapy and strength training, gradually rebuilding stamina, and mobility. As pain decreased, energy and confidence returned, making later professional adaptation far more sustainable. At the plant level, physical adaptation after the incident involved upgrading personal protective equipment (PPE), redesigning access routes, reducing unnecessary climbs and lifts, and rotating staff away from the most fatiguing tasks. When physical adaptation is taken seriously, bodies are no longer treated as passive carriers of resilience but as active components of it. Muscles, joints, and nervous systems are given the chance to recover, tools and layouts are adjusted to fit human limits, and the risk of secondary injuries during the vulnerability period is reduced. In turn, this physical stability supports emotional and mental adaptation, making it easier to think clearly, regulate feelings, and engage with the long climb of recovery.

Digital Adaptation

In the digital age, information overload can intensify vulnerability. Adaptation therefore involves setting clear boundaries on digital consumption and using technology as an aid to recovery rather than a source of additional strain. During one vulnerability period, news intake was deliberately limited to a fixed window each day, simple productivity tools were used to organise tasks, track small wins, and reduce cognitive clutter. At the plant, digital adaptation took a different form. Secure channels were established for sharing real-time safety data, and dashboards were developed to monitor compressor performance and flag emerging issues before they escalated. Technology shifted from being a noisy background presence to a focused support for decision-making. Digital adaptation also includes a willingness to learn new tools. During the pandemic, many teams were forced to move rapidly onto remote collaboration platforms, digital whiteboards, and online training environments. Those who adapted most effectively were not necessarily the most technically gifted, but those who treated digital tools as extensions of resilience: ways to stay connected, coordinate safely, and keep learning even when the usual physical structures disappeared.

Social Adaptation

Social connections sit at the heart of resilience. The Resilience Activation Framework suggests that access to social resources, formal and informal support, community services, and spiritual guidance, can trigger and sustain resilience processes. Social adaptation may involve reaching out to trusted friends and mentors, joining support groups, and quietly redefining relationships that no longer serve recovery. In one period of job loss, shame had to be faced before a mentor could be told the full story; his honest guidance then became a turning point, sharpening the focus of the job search and restoring a sense of direction. At the plant, daily briefings were used not only to coordinate tasks, but also to hold morale, share fears, and remind the team that no one was carrying the incident alone. Social adaptation also means negotiating boundaries. Some connections nourish recovery; others quietly erode it. Part of resilience is learning to limit contact with those who drain energy or undermine confidence, deliberately investing time in relationships that stabilise, strengthen, and support the climb back from disruption.

Professional Adaptation

Professionally, adaptation often means altering work practices and, at times, redefining one's role altogether. After a sudden job loss, adaptation might involve taking on freelance projects, enrolling in targeted courses, and offering safety expertise to smaller organisations that would otherwise lack access to that knowledge. At the plant level, professional adaptation can be seen in the reassignment of staff, the retraining of operators, and the engagement of external consultants to close critical gaps. In the wider energy industry, it may take the form of shifting towards renewable sources, redesigning assets, or diversifying revenue streams to reduce dependence on a single technology or market. On a more personal scale, adaptation can mean updating a résumé, learning new software, or accepting a different level of responsibility than before. The difficult, quiet skill in all of this is knowing when to pivot and when to persevere, when to hold a course that still has life in it, and when to step into a new direction so that growth can continue.

A Lived Adaptation Story – the Gas-Turbine Failure

My deepest lessons about adaptation came during a three-month period as a work health and safety advisor at a power plant. On a humid summer afternoon, a gas turbine experienced a catastrophic blade failure. Metal fragments tore through the casing, causing a fire and threatening nearby equipment. The shock was immediate: alarms blared, diesel generators kicked in, and we had minutes to prevent a secondary explosion. Absorption meant shutting down adjacent systems and evacuating personnel; we contained the physical hazard and accounted for everyone. Once the flames were out and the area secured, PF plummeted, our production dropped to zero, and we had no clear path forward. We entered the vulnerability period.

In the first 24 hours, adaptation became my primary mission. We needed to restore some capacity quickly and prevent new hazards. We convened our technical experts and examined sensor data and maintenance logs to understand what happened (sense-making). We then identified our priority: restore partial power by restarting unaffected turbines. Non-critical maintenance tasks were deferred, and the workforce was concentrated on the failure analysis (prioritisation and resource reallocation). We formed an MVR: each day began with a morning technical briefing, followed by safety walk-throughs, midday progress review, and evening debrief. We instituted guardrails: no modifications without peer review and strict overtime limits. We emphasised redundancy by sourcing spare parts from sister plants and verifying backup generators. Importantly, we reframed the event for our team: instead of labelling it a failure, we called it a stress test of our resilience. These adaptation steps slowed the decline and nudged our PF upward. Within a week, we had restored 50% output. The vulnerability period was far from over, we were still operating below baseline and could easily suffer another setback, adaptation gave us a foothold.

RECOVERY: THE CLIMB BACK TO BASELINE

What Recovery Means

If adaptation is the pivot, recovery is the climb. Recovery refers to the process of rebuilding or restoring lost capacity after a disruption. It is a deliberate, often gradual movement towards the desired level of performance. Engineering definitions of resilience emphasise rapid recovery, yet scholars caution that rapid return to pre-impact conditions can be problematic if pre-existing vulnerabilities are unaddressed. Instead, resilient recovery involves both regaining what was lost and integrating lessons, and reducing future vulnerabilities. The Systems Engineering Body of Knowledge (SEBoK) taxonomy lists recovery as one of the fundamental objectives of resilience, alongside avoiding and withstanding adversity. Recovery is distinct from adaptation: adaptation modifies operations to function in the present; recovery increases capacity over time until the gap between current performance and baseline shrinks. In the Wither or Thrive model, the recovery trajectory describes a decline in functioning followed by improvement, whereas the resilience trajectory maintains function throughout. Recognising this distinction prevents us from conflating short-term adjustment with long-term healing.

One way to think about recovery is through the lens of PF slope and PF threshold. Adaptation flattens the decline and begins to raise the PF; recovery steepens the slope and aims to reach the threshold of normal operation. Recovery often takes longer than adaptation because restoring strength, whether physical, emotional, organisational, or infrastructural, requires time, resources, and deliberate effort. Rapid improvement can occur; plateaus and relapses are common. The goal is not to replicate exactly what was before, it is to regain function which allows for meaningful activity and quality of life. In some cases, the new normal may be different: a company may choose a leaner operational model, a person may shift to a different career, a community may rebuild with new infrastructure. Recovery is successful when PF is sustainably high, and vulnerability is low.

Principles of Effective Recovery

Several principles guide effective recovery:

1 *Define a target baseline (old or new).* Before you can close the gap, you need to know where you are heading. Decide whether you aim to return to the old standard (e.g., restore production to previous levels), or to create a new baseline (e.g., move to a hybrid workforce). Recovery may involve choosing a different business model or lifestyle after the disruption.
2 *Sequence capacity blocks.* Recovery works best when you rebuild capacity in a logical order. For individuals, focus first on basic needs (sleep, nutrition, physical-therapy), then on cognitive functions

(focus, learning), then on complex tasks (work projects). For organisations, start with safety and quality, then throughput, then innovation. At our plant, we first restored safe operations, then increased output, then resumed deferred maintenance and improvement projects.

3 *Use staircase progression.* Recovery is not linear; you can accelerate progress by alternating periods of exertion and consolidation. In physical rehabilitation, this might mean two weeks of increasing load followed by a deload week. In professional recovery, it might mean gradually increasing workloads, and scheduling recovery days. The key is to avoid continuous push, which can lead to burnout or re-injury.

4 *Gate progress on criteria, not calendar.* Resist the urge to advance on a fixed timeline. Instead, set gates, performance criteria that must be met before scaling. Example gates might include three consecutive weeks of stable sleep, or two weeks of error rates below a threshold, or a certain amount of cash reserves. At the plant, we mandated that no capacity ramp could occur until we had ten consecutive shifts without safety incidents.

5 *Monitor relapse response time.* Relapses will occur, emotional triggers, equipment hiccups, and funding shortfalls. The measure of resilience is not zero setbacks, it is how quickly you return to your recovery trajectory. Track how long it takes to bounce back after a wobble; decreasing response time is a sign of progress. Plan ahead for setbacks and have contingency plans ready.

6 *Integrate adaptation and recovery.* Adaptation does not end when recovery begins. In many cases, the return to normal operations exposes new points of strain. As production increased at the plant, it became clear that the supply chain no longer had sufficient slack; additional redundancy was needed, and a second spare-parts supplier was contracted to prevent future bottlenecks. In a similar way, once full employment resumed, work habits had to be adjusted to include regular learning and upskilling, so that professional competence did not quietly drift towards obsolescence. Recovery, in this sense, is not a finish line but a new stage in which fresh adaptations are required to keep the system strong.

Recovery Across Domains

Recovery is multi-dimensional; it unfolds across the same domains in which adaptation occurred.

Emotional Recovery

Emotional recovery involves stabilising moods, processing grief or trauma, and gradually regaining a sense of safety and joy. Psychological research emphasises that positive emotions can broaden coping and accelerate stress recovery. In practice, emotional recovery may include therapy, peer support, mindfulness, gratitude practices, and re-exposure to previously avoided contexts. For me, emotional recovery after job loss included developing a regular journalling habit, practising compassion towards myself, and gradually reintroducing professional social interactions. At the plant, emotional recovery meant taking time to celebrate small milestones (e.g., a week without incidents) and acknowledging team members' efforts.

Mental Recovery

Mental recovery is the slow return of clarity, focus, and confidence after a shock. It rarely happens under constant pressure; it usually asks for rest, gentle learning, and a gradual increase in cognitive load. Following a period of job loss, a deliberate season of low mental demand was created: days were filled with walks, quiet reading, and simple tasks rather than complex decisions. When the mind began to feel less fogged and less raw, an online course was added to rebuild skills, and work commitments were increased step by step instead of all at once. At the plant, mental recovery took a more collective form. The incident was carefully debriefed, cognitive biases behind key misjudgements were named, and staff were retrained with those insights in mind. Engineers were encouraged to take real breaks, step away from control rooms, and avoid making critical decisions while exhausted. In both settings, the message was the same: the mind is an asset, not an endless resource, and recovery depends on giving it time, structure, and space to heal.

Physical Recovery

Physical recovery refers to the restoration of bodily strength and health. For individuals, this may involve physiotherapy, structured exercise programmes, careful attention to sleep, and sustaining nutrition that genuinely supports healing. For organisations, physical recovery can mean rebuilding infrastructure, repairing damaged equipment, or restoring supply chains so that operations can function safely again. After the turbine failure, the damaged unit was repaired, worn components across the plant were replaced, and fire suppression systems were upgraded to a higher standard. On a more personal level, a period of unemployment became an opportunity to address long-standing lower back pain through physiotherapy and strength training. As pain eased and movement improved, energy levels rose and overall resilience deepened, showing how closely physical repair and adaptive capacity are linked.

Digital Recovery

Digital recovery means reshaping digital tools and habits so they support, rather than erode, sustainable performance. It often involves curating a small set of trustworthy information sources, automating repetitive tasks, backing up critical data, and streamlining cluttered workflows. At the plant, this took the form of simplified digital reporting and the creation of a central data repository for maintenance records, reducing confusion and duplication. On the personal side, email inboxes were cleared, distracting newsletters were abandoned, and a simple digital system was built to track job applications and follow-ups. In this way, technology is brought back into its proper role: not as a constant demand on attention, but as a quiet, reliable ally in recovery and ongoing work.

Social Recovery

Social recovery involves rebuilding and strengthening relationships and networks after disruption. Work on the Resilience Activation Framework highlights that access to social resources, friends, colleagues, mentors, communities, and support services can activate resilience processes and significantly improve adaptation and recovery. When work is lost or identity feels shaken, renewed attention to connection becomes essential rather than optional. In one such period, neglected friendships were repaired, professional connections were renewed, and involvement in a local community group created a new sense of belonging and contribution. At the plant, social recovery meant restoring trust between management and frontline workers, hosting an open day for families to witness the safety improvements first-hand, and arranging cross-functional workshops so teams could listen to one another and rebuild collaboration. Through these steps, the social fabric around the disruption grew stronger than it had been before, providing a living network of support for the next challenge.

Professional Recovery

Professional recovery entails regaining competence, productivity, and career momentum. For individuals, this may involve updating skills, re-entering the workforce, or pivoting to a new field. For organisations, it involves ramping up production, restoring service levels, and delivering on commitments. After the turbine incident, we gradually increased power output, cleared maintenance backlogs, and resumed improvement projects. On a personal level, my professional recovery involved completing new certifications, taking contract work in adjacent fields (renewable energy and safety auditing), and eventually securing a new full-time role that aligned with my values and leveraged my expanded skill set.

A Lived Recovery Story – Climbing from Adaptation to Baseline

The adaptation steps after the gas-turbine failure set the stage for recovery, but the climb itself required months of disciplined effort. Once we had restored 50% output, our challenge was to return to full capacity without replicating the conditions that led to the failure. We defined our target baseline not as the old 100% output but as 100% output with enhanced safety and reliability. We sequenced our recovery: first we repaired and tested the damaged turbine (physical recovery), then we retrained operators and updated procedures (mental and professional recovery), and finally we resumed non-critical maintenance and improvement projects (capacity building). We used a staircase progression: each week we increased output by 10%, followed by a review and a partial deload to consolidate. We established gates: no ramp beyond 70% unless we had 14 consecutive days without safety incidents, and no final ramp to 100% until the root-cause analysis of the failure was peer-reviewed and addressed. We monitored relapse response time by tracking how quickly we returned to stable operation after minor glitches; as weeks went by, our response time shortened. We integrated adaptation into recovery by continuing to use the MVR, guardrails, and redundancy planning throughout the climb.

In personal recovery, the climb looked different, yet it followed the same underlying principles. After months of adaptation, establishing a new routine, reshaping identity, and building fresh skills, a quiet readiness to regain full professional capacity began to emerge. A target baseline was defined: to be fully employed in a role that drew on safety and risk expertise, while creating space to learn and contribute to renewable energy work. Capacity was rebuilt in stages. First came physical and emotional health: consistent sleep, regular movement, and therapeutic support to steady the nervous system. Next came mental focus: structured learning, reading, and deliberate practice to sharpen concentration again. Only then did professional action take centre stage, networking, interviews, and short-term contracts. A staircase approach was used, gradually increasing the number of job applications and weekly working hours rather than leaping straight back into a full load. Clear gates were set to prevent a rushed return. A full-time role would only be accepted once a relevant certification had been completed and emergency savings covered at least six months of expenses. When setbacks arrived, a failed interview, or a family illness, the measure of progress became the time it took to regain focus. Slowly, the "bounce back" shortened; shock no longer held attention hostage for as long. After six months, a position with a renewable energy company was accepted, marking the end of that vulnerability period. The role was not a replica of the previous job. It drew on a familiar safety background but demanded new technical knowledge and a broader outlook. This new baseline felt closer to core values and less exposed to the volatility of earlier industrial sectors, a different kind of stability, built on intentional recovery rather than habit alone.

ADAPTATION VS RECOVERY: A COMPARATIVE LENS

Understanding the difference between adaptation and recovery is critical because the interventions, metrics, and mindsets for each are different. Adaptation is about continuing to function under new constraints; it focuses on process modifications, resource reallocation, and psychological reframing. Recovery is about regaining capacity; it focuses on restoring strength, rebuilding infrastructure, and returning to (or redefining) baseline. They overlap in time, as adaptation begins first, continues during recovery, and influences the quality of recovery, but they are conceptually distinct.

In the following table, there are some key differences:

ASPECTS	*ADAPTATIONS*	*RECOVERY*
Purpose	Maintain function under changed conditions	Regain lost capacity
Timing	Begins immediately after absorption	Begins after adaptation has stabilised function, continues until baseline is reached
Focus	Process adjustments, resource pivots, narrative reframing	Capacity building, healing, infrastructure repair, skill restoration
Metrics	PF floor, error rate, energy budget, adaptation actions	PF slope, load tolerance, relapse response time, progress against gates
Mechanisms	MVR, guardrails, redundancy, sense-making, prioritisation, social support activation	Sequenced capacity blocks, staircase progression, gates, deliberate practice, skill development
Outcome	New operational norm that works within constraints	Return to baseline or creation of a new baseline with reduced vulnerability

Seeing these differences helps practitioners and individuals allocate their efforts appropriately. In the early stages of the vulnerability period, adaptation actions should dominate: gather information, stabilise routines, reallocate resources, and set boundaries. Once adaptation has created stability, recovery actions can increase: rebuild capacity, engage in therapy or training, and remove temporary constraints. Recognising the difference prevents premature recovery attempts that ignore unaddressed adaptations, which can lead to relapse or re-injury. As the Resilience Activation Framework reminds us, failing to identify and address resource deficits can lead to rapid loss cycles and worsening mental health outcomes. In other words, if you try to climb back without pivoting your footing, you might slip again.

RESEARCH PERSPECTIVES ACROSS DISCIPLINES

Community Resilience and Disaster Management

Disaster and community resilience research has long grappled with the interplay of adaptation and recovery. The 2012 U.S. Presidential Policy Directive on critical infrastructure security and resilience defined resilience as the ability to prepare for, adapt to, withstand, and rapidly recover from disruptions, emphasising adaptation alongside recovery. A comprehensive review of community resilience literature highlighted that resilience is both resistance to disruption or rapid recovery and includes learning and adaptation to future shocks and vulnerabilities. This review cautioned that a return to pre-impact conditions without addressing vulnerabilities can be problematic. These insights prompted policymakers to integrate adaptation into disaster cycles. The NATF/EPRI definition we discussed earlier similarly emphasises adaptation and recovery as separate, overlapping components.

The COVID-19 pandemic led to calls for adding an adapt stage to the disaster management cycle. A structural-engineering perspective argued that communities remained in prolonged duress, oscillating between response and recovery; an explicit adapt phase would allow for rethinking infrastructure and reducing vulnerability. The same authors noted that simply rebuilding to pre-pandemic conditions would leave societies vulnerable to future shocks and advocated for innovative approaches to recovery. These proposals echo experiences from previous disasters: after hurricanes, communities that rebuilt to stronger building codes and diversified economies fared better than those that restored identical infrastructure. Access to resources such as social networks, economic reserves, and governance capacity can enable adaptation and accelerate recovery, highlighting the importance of equity and resource distribution in resilience planning.

Psychological and Neuroscientific Research

Psychologists examine adaptation and recovery at the level of individuals. Research shows that resilience is a dynamic process influenced by multiple independent factors, personal traits, environmental context, and stressor characteristics, which interact over time. The distinction between resilience factors (trait-like predispositions such as optimism or mastery) and resilience mechanisms (processes such as cognitive reappraisal or social support) helps clarify how adaptation operates. Some people possess traits that predispose them to adapt quickly; however, everyone can engage in mechanisms which facilitate adaptation and recovery. The Wither or Thrive model emphasises that resilience trajectories vary; some individuals maintain function (resilience trajectory), others experience an initial drop and then recover (recovery trajectory), and still others show chronic dysfunction or delayed recovery. This model points to the need for tailored interventions.

Another recent paper on psychological resilience proposed a transversal definition: resilience is the ability to recover from adverse or changing situations through a dynamic process of adaptation, influenced by personal characteristics, family, and social resources. The authors reviewed numerous definitions and found consensus that resilience involves both recovery and positive adaptation. They compiled a table of definitions emphasising the capacity to rebound, adapt, and sometimes even thrive. Such reviews highlight that resilience is not static; it is a process that evolves through adaptation and recovery.

Neurological research adds another layer. Studies on stress physiology show that resilience is associated with efficient regulation of the stress response. After a shock, the body's cortisol and adrenaline spikes subside, but in resilient individuals, neural circuits for cognitive control and emotional regulation become more active, enabling adaptation. This aligns with the Resilience Activation Framework, which emphasises that adaptation involves the mobilisation of human capital (health, coping skills, and emotional regulation). Adaptive coping strategies such as problem-solving, positive reframing, and seeking social support have been shown to buffer stress and promote faster recovery.

Engineering and Systems Resilience

In engineering and systems disciplines, resilience is framed around maintaining and restoring function. The SEBoK definition states that resilience includes the ability to avoid, withstand, and recover from adversity, and that adaptation is a means-objective that allows systems to continue delivering capability under changing conditions. Engineers model resilience curves showing rapid drops, adaptation phases, and recovery slopes. They emphasise design for flexibility (e.g., modular systems, redundancies), rapid reconfiguration, and planned recovery protocols. For example, after the Fukushima nuclear accident, Japanese engineers redesigned cooling systems to operate without power and implemented mobile backup units, an adaptation that allowed for faster recovery in later emergencies.

Infrastructure resilience considers time to recovery and damage avoided. Adaptation reduces damage by modifying operations during the hazard, while recovery focuses on repairing damage. By tracking both metrics, engineers can prioritise investments. For instance, in power grids, a resilience strategy might include early detection (sensors), adaptive rerouting (switching to alternate lines), and rapid repair crews. Research emphasises that adaptation and recovery should be planned together, with adaptation informing recovery priorities. The NATF/EPRI document clarifies that recovery activities begin almost immediately after an event and overlap with absorption and adaptation. In practice, this means that as soon as the hazard is contained, crews start assessing damage and planning repairs, while others reconfigure operations to maintain service.

Public Health and Community Psychology

Public health studies focus on the health impacts of disasters and the role of social resources. The Resilience Activation Framework emphasises that access to social resources such as formal and informal support can promote positive adaptation and rapid recovery. It identifies multiple forms of capital: human (health, coping skills), social (networks, support), economic (savings, insurance) and political (trust in governance), that interact to influence resilience. The framework also notes that persistent resource loss can lead to a downward spiral of vulnerability and mental health problems. Recognising adaptation and recovery as resource-dependent processes, public health approaches emphasise the need for equitable resource distribution and support systems. Studies of communities affected by hurricanes, oil spills, and pandemics show that those with strong social cohesion and access to mental health services adapt more quickly and recover better.

PRACTICAL GUIDANCE: DESIGNING YOUR VULNERABILITY BLUEPRINT

The theory and stories presented so far offer insight into adaptation and recovery. Nonetheless, how can you apply these ideas in your own life or organisation? The following framework synthesises key principles into a practical blueprint. Use it to map your vulnerability period, plan adaptation and recovery actions, and monitor progress.

Map Your Shock and Absorption

1. *Identify the shock.* Begin by naming what happened – the event that disrupted your life or organisation. Be specific and concrete. For example: "Laid off from my role on 15 March 2024", or "Turbine failure in Unit 3 at 14:52 on 3 July 2023". Writing down the exact event, time, and context draws a clear line on the timeline and marks the beginning of the vulnerability period.
2. *Describe the absorption.* List the actions you took to contain the immediate damage. Did you evacuate, call emergency services, or secure finances? Note how long the absorption phase lasted and what resources were consumed. Understanding your initial response provides context for adaptation.

Plan Your Adaptation

1. *Clarify constraints.* Determine what has changed in your environment, what resources are limited, and what remains stable. Ask yourself: What is no longer possible? What new opportunities have arisen? This forms the basis for sense-making.
2. *Set an MVR.* Create a daily or weekly routine that ensures you sleep, eat, move, connect, and do a minimum amount of high-value work. Keep it simple. In my job loss, my MVR was four blocks: two job applications, one hour of exercise, one hour of learning, and one meaningful conversation per day. At the plant, our MVR was four meetings and two checks per day.
3. *Identify adaptation levers.* Which mechanisms (sense-making, prioritisation, resource reallocation, guardrails, redundancy, narrative reframing) are most relevant? For each lever, write a specific action.

For example, "Reallocate $5,000 from marketing to safety training", or "Set a rule: no email responses after 8 p.m."

4 *Activate social resources.* List people and institutions you can lean on: mentors, colleagues, family, community groups, professional networks, and therapists. Reach out proactively. According to the Resilience Activation Framework, social resources can activate resilience processes and facilitate adaptation and recovery.

5 *Reframe your narrative.* Write a sentence that deliberately frames the disruption as a learning process rather than a personal failure. For example, instead of "I failed", use language such as "I am in a learning phase that will make me a more effective professional" or "This setback is teaching me skills I will use for years". Keep this statement where it can be seen often and use it as an anchor for your adaptation efforts.

Design Your Recovery

1 *Choose your target baseline.* Decide whether you want to return to your previous level of performance or create a new baseline. If your old job was unstable or unsatisfying, you may aim for a different role or lifestyle.

2 *Sequence capacity blocks.* Break down your recovery into stages and order them logically. Start with foundational needs (rest, health), then cognitive and skill development, then productivity and innovation. For organisations, this might mean safety repairs → core operations → capacity expansion → improvement projects.

3 *Plan staircase progression.* Set incremental goals with built-in consolidation periods. For example, increase workload by 10% each week and schedule rest every fourth week. In organisations, ramp up output gradually and include pauses to evaluate systems.

4 *Set gates.* Define specific criteria that must be met before moving to the next stage. Use objective metrics: number of error-free days, financial reserves, completion of training, and physical health markers. Resist the temptation to advance prematurely.

5 *Prepare for setbacks.* Expect relapses and define a response plan. If you experience a setback, review which adaptation lever or recovery block was insufficient and adjust accordingly. Measure how quickly you return to baseline after the setback; decreasing response time is a sign of improvement.

6 *Integrate ongoing adaptation.* Recognise that adaptation continues throughout recovery. New challenges will emerge as capacity increases. Maintain the MVR and guardrails until you are confident

that recovery is stable. Use adaptation mechanisms to address emerging constraints.

Quick Practice: Mapping Your Adaptation and Recovery

To make these concepts tangible, try this short exercise. Use this practice to reflect on a recent disruption and design your vulnerability blueprint.

1 *Shock and absorption.* Write one sentence describing the disruption. Then list the actions you took to contain it. How long did the absorption last?
2 *Adaptation planning.* Identify three constraints you face now. For each, note one adaptation lever you can use. Write your MVR for the next seven days.
3 *Recovery design.* Define your target baseline. List three capacity blocks (e.g., rest, learning, and work). For each block, set a staircase goal and a gate. For example, "Sleep: aim for seven hours nightly; gate: maintain seven hours for two weeks before adding extra workload".
4 *Social activation.* Write down two people you will contact this week for support or mentorship. Plan your outreach.
5 *Narrative reframing.* Write a one-sentence story about your disruption that emphasises growth ("This experience is teaching me . . .") and place it somewhere visible.

This practice is not about achieving perfection; it is about creating structure in uncertainty. Each iteration builds resilience by making adaptation and recovery conscious and intentional. Over time, as you repeat the exercise with different disruptions, you will notice patterns and refine your approach.

CLOSING REFLECTION – VULNERABILITY AS AN OPPORTUNITY

The vulnerability period is uncomfortable. It exposes cracks in your systems and can make you feel fragile, uncertain, and impatient. Yet it is also fertile ground for growth. By distinguishing between adaptation and recovery, we gain tools to navigate this period with intention. Adaptation is the agile pivot that keeps you operating within constraints. Recovery is the disciplined climb back to strength. Together they allow you to cross the gap between the shock and the new normal. Without adaptation, recovery is precarious; without recovery, adaptation stagnates.

Recent research from engineering, psychology, and public health reinforces these insights. For example, community resilience studies highlight that resilience is resistance, rapid recovery, and adaptation. Models like the Wither or Thrive framework distinguish between resilient trajectories and recovery trajectories and show how individual and environmental factors shape these paths. The Resilience Activation Framework reminds us that access to social, economic, and HR can activate adaptation and speed recovery. Psychological definitions emphasise that resilience is a dynamic process of recovering through adaptation. Engineering guidelines call for adaptation and recovery to be planned together, with adaptation informing recovery priorities. These perspectives converge on the idea that resilience is a journey across a shifting landscape, not a static trait.

The experiences traced here, from turbine failures to career upheavals, show that the vulnerability period is both perilous and potentially transformative. In each case, the turning point arrived when adaptation was treated as an active discipline rather than a passive wait: data were gathered, routines were established, resources were reallocated, narratives were reframed, and recovery was then pursued with intention by sequencing capacity, setting gates, and measuring progress. Lessons drawn from these exposed periods do not fade; they become part of a resilience toolkit. They build a quiet confidence that when the next shock comes, it is possible to pivot, to climb, and to emerge stronger. By deliberately designing a vulnerability blueprint and practising adaptation and recovery, a person or organisation moves from being at the mercy of events to becoming, step by step, an architect of their own resilience.

PRACTICE WORK: APPLYING ADAPTATION AND RECOVERY IN YOUR PROFESSIONAL CONTEXT

This section is designed to help you translate theory into action. The goal is to make the concepts of adaptation and recovery tangible so you can apply them in your current work environment, whether you are an individual contributor, a team leader, or an organisational decision-maker.

Step 1: Reflect on Your Disruption

Start by identifying a recent or ongoing disruption in your professional life. This could be:

- A sudden job loss or role change
- A major project failure
- A shift in organisational structure
- A personal crisis that impacts your work

Questions to guide you:

- What exactly happened? Be specific.
- How did this event affect your PF – emotionally, mentally, physically, or professionally?
- What immediate actions did you take to absorb the shock?

Step 2: Map Your Vulnerability Period

The vulnerability period is the time after the shock when you are no longer in freefall but are not yet stable. Describe this phase:

- What constraints did you face? (e.g., reduced resources, unclear priorities, emotional fatigue)
- What remained stable or supportive during this time?
- How long did this period last, and did any secondary shocks occur?

Step 3: Design Your Adaptation Strategy

Adaptation is the pivot under load, the set of actions that allow you to function under new conditions.

Action prompts

- Identify three constraints you are currently facing.
- For each constraint, choose one adaptation lever (e.g., sense-making, prioritisation, resource reallocation, guardrails, redundancy, or narrative reframing).
- Write a specific action for each lever.

Example: "Reallocate 10% of my time from meetings to strategic planning".

Create your MVR

Define a simple daily or weekly structure that ensures:

- Sleep and rest
- Learning or skill-building
- Connection with others
- Core work tasks
- Keep it realistic and repeatable.

Step 4: Activate Social Resources

Resilience thrives on relationships. List the people, networks, and institutions you can draw upon:

- Who can offer mentorship, emotional support, or professional guidance?
- What communities or groups can you engage with?
- Plan two outreach actions this week (e.g., schedule a call with a mentor, attend a peer support meeting).

Step 5: Reframe Your Narrative

The story you tell yourself shapes your recovery trajectory.

- Write a one-sentence reframing of your disruption
  - *Instead of I failed, try This experience is teaching me to lead with clarity and compassion.*
- Place this statement somewhere visible in your workspace.

Step 6: Build Your Recovery Blueprint

Recovery is the climb back to baseline, or to a new, more resilient standard.

Define your path

- Choose your target baseline: Are you aiming to return to your previous level or create a new one?
- Break recovery into capacity blocks:

- Foundational needs (health, rest)
- Cognitive development (learning, focus)
- Productivity (projects, innovation)
- For each block:
- Set a staircase goal (incremental progress).
- Define a gate (criteria to move forward).

Example: "Health: Walk 30 minutes daily; gate: maintain for 3 weeks before increasing intensity".

Step 7: Monitor and Adjust

Recovery is not linear. Expect setbacks and plan for them:

- Track your relapse response time: How quickly do you return to your recovery path after a disruption?
- Adjust your adaptation levers and recovery blocks as needed.
- Celebrate small wins and reflect on progress weekly.

Step 8: Apply in Team or Organisational Settings

If you lead others, facilitate a **resilience workshop**:

- Ask team members to identify current constraints.
- Choose one adaptation lever to implement collectively.
- Design a team MVR and set recovery gates.
- Use narrative reframing to foster a culture of learning.

Closing Thought

Engaging deeply with these exercises, you transform abstract concepts into practical tools. Adaptation and recovery are not just responses to crisis, they are ongoing practices that build resilience, foster growth, and prepare you for future challenges.

REFERENCES

1. Platt, S., Brown, D. and Hughes, M., 2016. Measuring resilience and recovery. *International Journal of Disaster Risk Reduction*, 19, pp.447–460.
2. Fletcher, D. and Sarkar, M., 2013. Psychological resilience. *European Psychologist*, 18(1). https://doi.org/10.1027/1016-9040/a000124
3. Dahles, H. and Susilowati, T.P., 2015. Business resilience in times of growth and crisis. *Annals of Tourism Research*, 51, pp.34–50.
4. Yazdi, M., 2024. Reliability-centered design and system resilience. In *Advances in computational mathematics for industrial system reliability and maintainability*. Cham: Springer Nature Switzerland, pp.79–103.
5. Godara, M., Silveira, S., Matthäus, H. and Singer, T., 2022. The wither or thrive model of resilience: An integrative framework of dynamic vulnerability and resilience in the face of repeated stressors during the COVID-19 pandemic. *Adversity and Resilience Science*, 3(4), pp.261–282.

Learning *Integrating Lessons for Future Resilience* 5

OPENING REFLECTION, STEPPING ONTO NEW GROUND

For last year's words belong to last year's language, and next year's words await another voice.

T. S. Eliot

The ridge stretched out, bathed in the amber glow of late afternoon. Months had passed since the day belongings were carried out of that office, heart pounding with uncertainty. Now, on this windswept hilltop trail, the quiet weight of perspective settled in. Below lay the valley of all that had once felt secure: the job titles, the routines, and the safety nets taken for granted. Ahead, the path wound forward into uncharted terrain. The person who stumbled into the harsh light of sudden upheaval no longer existed. Panic and grief had given way to curiosity, then determination, and finally insight. Each setback had carved new understanding: about strengths, values, and the kind of life worth rebuilding. Fear, once knotted tight, was tempered by hard-earned confidence, a belief that learning could come from whatever lay ahead. That confidence carried a deeper realisation: the difference between living for utility and living for status. The old life had been filled with symbols, titles, logos, and prestige, which promised meaning but often delivered little more than validation from others. True value, it turned out, was not in appearances but in what made life genuinely better: comfort over display, function over brand, and substance over show. Like choosing a well-equipped Toyota over a bare-bones Bayerische

DOI: 10.1201/9781003789895-6

Motoren Werke (BMW, which translates from German to English as Bavarian Motor Works), the measure was no longer what impressed others but what sustained joy and purpose when no one was watching. On this trail, with the horizon wide open, the choice was clear. The next chapter would not be built on status but on utility, on things, habits, and work that mattered for their own sake. A life owned for oneself, not for the gaze of others.

Not long ago, mornings were spent poring over incident reports and engineering diagrams, dissecting failures to prevent future ones. Now, the same process applies to lived experience, replaying critical moments, asking what really happened, and why. On this ridge, the past year unfolds like a story read somewhere: the professional loss that landed like a punch, the slow climb through freelance work and side projects, the mentors and friends who appeared like handholds, and the unexpected opportunity that finally clicked into place. Each chapter carried its own lesson. Losing a job revealed which parts of identity could survive without a business card. Learning new skills on the fly proved adaptability, when pride gave way to curiosity. Asking for help showed that resilience is woven through community, not a solitary act of heroism. And beneath those lessons lies something deeper: the importance of escaping the cultural bubble that equates worth with status. Without that escape, propaganda wins, and modern delusions take root. True wealth is not measured by titles or possessions but by autonomy, the ability to remain grounded in values regardless of circumstance. When identity depends on external validation, ambition becomes a master rather than a tool. When satisfaction comes from within, money and status lose their power to dictate life. The ridge makes this clear: happiness cannot be outsourced to pay checks or prestige. It grows from character, relationships, and meaning, the things that endure when the logos fade. Pride belongs not to consumption but to what has been built: the friendships, the wisdom, and the memories. Richness is easy to buy; wealth is harder. Wealth is freedom from insatiability, the quiet confidence that life is shaped by choice, not by the gaze of others [1].

The crisp air settles in the lungs as eyes close for a moment. There is a sense of completion, but not closure, because resilience is not a finish line, it is a horizon that moves as the traveller moves. Every step taken to recover and adapt has led to this point, yet the journey continues. Stability no longer depends on solid ground; it is carried within. The earth will quake again, inevitably, but preparation feels stronger now. The difference lies in an invisible toolkit of lessons strapped to the back. Life has become both classroom and laboratory, each experience dissected like an after-action review (AAR), wounds transmuted into wisdom. This moment on new ground is both ending and beginning: the close of one resilience cycle and the quiet start of another. The path stretches forward, and each stride bears more than weight, it carries experience refined into insight. And insight, unlike fear, is light enough to carry anywhere.

This chapter emerges from that reflective vantage point. It examines learning as the crucial final stage of the resilience cycle, the phase where we integrate hard-won lessons and convert recovery into lasting growth. In the sections that follow, we will explore how reflective learning fortifies resilience, how feedback loops drive continuous improvement, and why learning from disruption *and* from success are both essential. We will consider how individuals, organisations, and even whole societies learn from adversity and feed that knowledge forward. Drawing on resilience engineering principles, psychological theory, and lived experiences, we will see how the journey from recovery to true transformation is paved by learning. Resilience is more than surviving a disruption; it is bouncing back with insight and often finding a stronger path in life. By the end of this chapter, we will understand how resilience becomes an ongoing evolution, a cycle that turns adversity into wisdom and change into growth [2].

THE ROLE OF REFLECTIVE LEARNING IN RESILIENCE

Resilience as a Learning Cycle

Resilience is not a static trait one simply *has* or *doesn't have*, it is a dynamic process that unfolds over time. Central to this process is the capacity to draw lessons from what we go through. In fact, many modern definitions of resilience include an element of learning or insight. For example, one psychological framework defines emotional resilience as the ability to recover from disruption, adapt to change, and bounce back with insight, often leading to a renewed or strengthened life path. That insight component, the reflective learning after a struggle, is what ensures that we don't simply return to our old equilibrium but potentially move towards a better one. Each setback or stress can, in hindsight, become a lesson on how to cope more effectively next time or how to avoid similar pitfalls.

Resilience can be viewed as a cycle with distinct stages: preparing for disturbances, absorbing the shock, recovering and adapting, and finally *learning* from the experience. This final learning stage closes the loop and feeds into future preparedness. Engineers and disaster management experts have increasingly recognised that after "respond and recover", an additional step is needed: learn and improve. The United Kingdom's emergency management model, for instance, explicitly adds a "learn and improve" phase following response and recovery, underscoring that every incident should inform plans. In this view, resilience is a continuous loop rather than a one-time bounce-back, each disruption teaches the system how to better handle the next. The continuous improvement principle [3] in resilient organisations similarly holds that the job is not finished when normalcy returns; it is finished only when we have extracted insights from the event and applied them to strengthen our defences.

The field of resilience engineering makes this idea even more explicit. Pioneers of resilience engineering, such as Erik Hollnagel, argue that a system (whether an organisation or a community) is truly resilient only if it possesses four key capabilities: to respond, monitor, anticipate, *and learn*. Learning in this context means absorbing the lessons of both failures and successes to continually adapt. A resilient power grid, for example, does not just bounce back after a blackout, it investigates what happened and updates its design or procedures to reduce the chance of recurrence. A resilient person, likewise, does not just "get over" a personal crisis, they reflect on it, derive meaning, and incorporate that new understanding into their life going forward. Researchers have noted that resilience entails resisting and recovering from impacts and adapting and learning to reduce future vulnerabilities. In other words, the experience of adversity can leave you stronger and wiser than you were before.

Reflective Practice and Insight

Why is reflective learning so powerful in the context of resilience? Psychologists point to the way self-reflection can transform painful experiences into sources of personal growth. When we engage in structured reflection, asking ourselves what happened, what we felt, why we reacted as we did, and what we might do differently, we activate a kind of internal feedback loop that turns experience into insight. One theoretical model, the Systematic Self-Reflection Model of Resilience Strengthening, proposes that deliberately reflecting on one's initial responses to stress is a key mechanism for building resilience over time. According to this model, when you revisit a stressful event in your mind and analyse your reactions, you develop clearer insight into your own coping capacities, recognise the limits of those strategies, and begin to consider alternative approaches for the future. In essence, reflection shines a light on what you already did right (your existing strengths) and where you struggled (areas to improve), leading to conscious adjustments in how you handle difficulties.

The experience echoes a powerful truth: meaning often emerges from contrast. In the weeks following job loss, journalling became a nightly ritual, sometimes a furious outpouring of frustration, other times a calm dissection of small victories and setbacks. Each entry served as an AAR, revealing patterns that shaped recovery. Days spent reaching out to colleagues and friends lifted perspective; days spent in isolation deepened despair. Recognising that contrast was like finding a lever: connection was not weakness but strength. This insight led to building a support network, reframing adversity as a teacher rather than a tyrant. In psychological terms, this was meaning-making, transforming raw events into a narrative of growth and values. Resilience theory underscores this: interpretation matters. A setback viewed as permanent defeat breeds stagnation; reframed as a lesson, it becomes fuel for adaptation. Carol Dweck's research on growth mindset illustrates the same principle, those who believe abilities can evolve see challenges as opportunities, fostering persistence. Happiness, too, thrives on contrast. Just as it is said marvelled at the colours of a drab carpet after decades of blindness [4], the smallest progress can feel extraordinary against the backdrop of loss. The most mundane victories, sending a proposal, learning a new skill, become radiant when contrasted with uncertainty. A good life is not everything desired; it is enough to meet needs and a little extra to savour. When every want is met, appreciation fades. But when progress follows struggle, gratitude deepens. Resilience and happiness share this secret: both are born not from abundance, but from the gap between expectation and reality, and the choice to see that gap as possibility.

There is also a deeper, almost spiritual aspect to reflective learning in resilience. Survivors of great adversity often speak of how their perspective on life changed because of what they went through. Psychologists call this post-traumatic growth (PTG), the phenomenon where struggling with a crisis leads to positive changes in one's self-perception, relationships, and life philosophy. PTG does not negate the pain or loss, but it means that out of the struggle, the individual attains new understanding or appreciation. For example, someone who endures a serious illness might come away with a greater appreciation for life, a clearer sense of priorities, and perhaps a newfound purpose to help others. These transformations are essentially profound lessons learned at an emotional and existential level. Tedeschi and Calhoun, who first defined PTG, describe it as "positive psychological changes experienced as a result of the struggle with trauma or highly challenging situations" [5–6], including increased self-awareness, closer relationships, and discovering new possibilities in life. In short, the act of reflecting and making meaning out of hardship can lead to *transformative* resilience, not just bouncing back to where you were, but bouncing forward to a new level of personal development.

FEEDBACK LOOPS AND CONTINUOUS IMPROVEMENT

Learning in the context of resilience is not a one-off epiphany, it is an ongoing process of feedback and adjustment. In engineering and systems thinking, a feedback loop is a mechanism by which a system's output is fed back into the system as input, influencing future outputs. Simply put, it is how systems self-correct or evolve over time. In human terms, feedback loops can be thought of as the cycle of action → reflection → adjustment → new action. Every time we face a challenge (action) and then take time to reflect on how it went, we create the opportunity to adjust our strategy and respond better the next time (new action). This principle underlies the concept of continuous improvement, whether in personal habits, organisational processes, or complex engineering systems.

Organisations that excel at resilience often have formal processes to ensure feedback is captured after any significant event. A popular tool is the AAR, a structured debriefing method originally developed by the U.S. Army and now used widely in business, emergency management, and healthcare. An AAR is essentially a guided conversation asking: *What was supposed to happen? What actually happened? Why were there differences? What went well? What could be improved?* The aim is to uncover both successes and failures in order to learn from them. Importantly, AARs are conducted in a spirit of *no-blame*, encouraging honest discussion without fear of punishment, so that everyone feels safe to speak up about mistakes or concerns. By design, an AAR highlights what went well and why (so those practices can be sustained or replicated), as well as what could be improved and how. This systematic reflection turns every project or incident into a source of data for learning. Over time, the insights from repeated AARs feed a continuous improvement loop that makes the team or organisation more adaptable and resilient. In fact, resilience experts note that *resilience efforts are only as strong as what you learn from them.* Every disruption, drill, or near miss should trigger an AAR or similar review, a structured debrief where teams reflect on what happened, what went well, and what needs improvement. This process is essential for identifying weaknesses and reinforcing strengths so that each ordeal leaves the organisation better prepared for the next.

Consider a concrete example: a manufacturing company experiences an unexpected production line breakdown that halts operations for a day. A non-resilient organisation might fix the machine and simply push to resume work, treating the breakdown as a one-time blip. A resilient organisation, by contrast, will convene an AAR as soon as the crisis abates. In that debrief, they might discover that a particular component failed due to poor maintenance scheduling (a flaw in process) and that the team initially responded slowly because the escalation protocol was unclear (a flaw in communication). They will also note what *prevented* the situation from being even worse, perhaps a quick-thinking technician who performed a temporary patch, or an alternate supply that kept some output going. Documenting these points, the company can now adjust its maintenance schedules and clarify its emergency response procedures. The next time a disruption occurs, chances are it will be resolved faster, or possibly averted entirely, thanks to those lessons learned. In this way, continuous improvement is the engine of resilience: each challenge becomes an opportunity to strengthen the system.

The journey applied the same principle of incremental improvement. After launching a freelance consultancy post-layoff, a Friday reflection ritual emerged: two simple lists, "What went well this week?" and "What can be better next week?" A stripped-down AAR for a one-person operation. One week, under "went well", appeared a candid conversation with a client that prevented a major misunderstanding. Under "do better", poor time management surfaced as the culprit behind unnecessary stress. Writing these down created a feedback loop: replicate the candid communication with future clients, fix the time issue with a scheduling app, and blocked focus hours. Small adjustments compounded over time. The business stabilised, confidence grew, not because problems disappeared, but because each one became a lesson. This process mirrors a deeper truth: progress feels most meaningful when contrasted with struggle. Just as Michael May marvelled at a drab carpet after decades of blindness, small wins shine brightest against the backdrop of uncertainty. A good life is not everything desired; it is enough to meet needs and a little extra to savour. When every want is met, appreciation fades. But when growth follows hardship, gratitude deepens. Resilience and happiness share this secret: both thrive on contrast, and on the choice to turn setbacks into stepping stones.

In organisational settings, this kind of learning loop often requires a culture that values and rewards learning. High-reliability organisations (HROs), such as in aviation, nuclear power, or healthcare, are known for their relentless focus on learning from even the smallest deviations. In an HRO, a near-miss (an incident that *almost* caused harm) is treated as seriously as an actual failure, because it carries rich information about what could go wrong. Rather than chalking up a narrow escape to luck, HRO teams dissect it to find underlying issues and reinforce what went right. As one guidance on high reliability notes, *An HRO considers a near miss as evidence of a system vulnerability and regards it as an opportunity to improve the process.* In practice, that means staff are encouraged to report near-misses and errors, and those reports lead to concrete changes in procedures or training. This echoes a key HRO principle called Commitment to Resilience, in which teams continuously learn from errors and share successful strategies, ensuring that knowledge gained from both failures and successes is disseminated and used to enhance safety. The takeaway for any organisation is clear: to build resilience, create robust feedback loops at all levels. Capture lessons from every event, good or bad, and feed them back into your planning, training, and processes. Over time, this institutionalises an adaptive capacity that can handle surprises because it is always evolving.

Finally, it is worth noting that feedback loops need not be overly formal or bureaucratic. The essence is simply to pause, reflect, and adapt. Whether it's a family having a kitchen-table talk after a chaotic week ("How can we make next week smoother?"), a sports team reviewing game footage to fine-tune their play, or a community group debriefing after a storm clean-up effort, the pattern is the same. Resilient systems at every scale tend to have this habit ingrained. They *learn fast.* They treat each experience, especially the painful or unexpected ones, as teachable moments. They shorten the time between *failure* and *improvement.* In a world where change and disruption are constant, that is an incalculable advantage.

LEARNING FROM DISRUPTION VERSUS LEARNING FROM SUCCESS

When we think about learning from experience, we often default to the idea of learning from mistakes or failures. Every disaster or personal failure shouts its lessons in hindsight: *Don't do that again!* Indeed, much of classical resilience thinking, in fields like safety engineering or disaster response, has centred on analysing what went wrong to avoid it in the future. This is crucial, but it is only half the picture. The other, often neglected, half is learning from success. In a resilience context, this means paying attention not just to why things break, but also to why things *don't* break; understanding how and why we sometimes achieve good outcomes despite challenging conditions.

Learning from disruption (failure): There is no doubt that disruptions command our attention and demand analysis. Catastrophic failures, traumas, and crises force us to stop and take stock. In organisations, a major failure might trigger an investigation or a task force to identify root causes. In personal life, a failure, like a failed exam, a business collapse, or a broken relationship, often leads to soul-searching (*What went wrong? What could I have done differently?*). These analyses are valuable. They can be painful, as they require confronting uncomfortable truths, but they shine light on weaknesses that need strengthening. For example, after the COVID-19 pandemic upended supply chains and public health systems worldwide, countless reviews and commissions were established to learn from the failures exposed. Hospitals examined their surge capacity and communication breakdowns; companies rethought their dependence on single suppliers; and governments rewrote pandemic playbooks. The underlying goal was to capture the hard lessons of the disruption so that we would be less vulnerable next time. A common saying in emergency management is *Never let a crisis go to waste*, meaning, use the window after a disaster to push through improvements while awareness is high.

Early in a career, the sting of professional failure left a lasting mark. A safety audit project went awry; critical non-compliance was overlooked, leading to a serious incident. In the aftermath, every missed detail was examined. Two lessons emerged. First, intuition matters: when something feels off, silence is costly. A subtle concern about that facility had been ignored, and the price was high. Second, checklists and protocols are not static; they must evolve because new mistakes always appear. That failure hurt, but it reshaped practice: new checkpoints were added, and a culture of openness was championed, where even junior voices were heard. The failure became a teacher.

This underscores a broader truth: all behaviour, and all decisions, make sense with enough context. Hesitation to speak up was rooted in fear of overstepping. Reliance on outdated protocols came from trust in established systems. Understanding those drivers doesn't excuse the outcome, but it explains it. Just as spending habits reflect lived experience, professional missteps often mirror the pressures and assumptions of the moment. Growth begins when those patterns are surfaced and reframed, not as permanent flaws, but as lessons. Failure teaches best when curiosity replaces judgement and when the question shifts from "Why did this happen?" to "What shaped this choice, and how can it change?"

Learning from success: Interestingly, success can also teach, but we have to be even more intentional to hear those lessons because success is quiet. When things go right, there is a tendency to simply celebrate and move on, assuming that success will repeat. But resilience can be bolstered by dissecting success with the same rigour as failure. Why did a particular strategy work so well? What allowed us to beat the odds this time? Could we replicate those conditions deliberately? In the realm of patient safety and organisational reliability, this idea has been gaining traction through what's known as the Safety-II approach. Safety-II, in contrast to traditional "Safety-I" (which focuses on preventing things from going wrong), emphasises learning from when things go right. The rationale is that on any given day, for every incident that goes wrong in a hospital, there are a thousand procedures that went right. Those successful operations and near-misses carry information about system resilience, how staff adapted to challenges, how latent safeguards worked, etc. By studying and learning from work-as-done when it succeeds, organisations can understand the adaptive strategies that practitioners use and reinforce them. In fact, one recent review highlights a growing number of tools designed to capture and learn from daily successes in healthcare, not just errors. This is a paradigm shift: moving from an exclusive focus on "What failed?" to additionally asking "What prevented failure?" and "What led to success under tough conditions?"

A practical example of learning from success comes from the wildland firefighting community. Firefighters, historically, would investigate tragic fire entrapments to learn how to avoid future fatalities (learning from failure). But they now also study "near-miss" fires where crews *survived against the odds.* One famous case involved a crew that correctly read a sudden weather change and withdrew just before a fire blew up. That wasn't an accident, it was the result of experience and intuition. By documenting how that crew sensed danger (they noticed subtle changes in smoke behaviour and wind direction), other crews were taught to heed those same signs. In essence, the success story, a disaster averted, became a lesson that potentially saved lives later. This approach requires humility and vigilance: humility to acknowledge that even our successes may involve a bit of luck or extra effort worth understanding, and vigilance to capture those fleeting insights.

Research in cognitive psychology suggests that positive outcomes can yield durable learning, sometimes even more so than negative outcomes, because success can reinforce effective behaviour if we consciously recognise what we did right. For instance, a student who aces an exam should reflect: what study methods worked well? Can I apply them again? Without that reflection, the student might attribute the success to innate ability or easy material and fail to carry forward any intentional strategy. Similarly, in business, a project delivered on time and under budget is a goldmine of what-to-do-right, if only teams routinely dissected success. Unfortunately, success often breeds complacency, the attitude that "if it ainot broke, do not fix it". Resilience thinking argues the opposite: *if it ain't broke, study it carefully, because something kept it from breaking.*

Balancing both: The most resilient individuals and organisations cultivate a learning mindset that is agnostic to outcome. Everything is information. A failure tells you one set of things; a success tells you another. Taken together, they provide a more complete picture. This balanced learning is mirrored in Hollnagel's four abilities of resilience (anticipate, monitor, respond, and learn) where *learning from experience requires actual events from both what goes well and what goes wrong*, not just accumulating data on failures. Embracing this can be a challenge culturally. Many organisations have robust incident reporting systems for errors but lack any mechanism to report "what went right". One technique borrowed from the military and aviation is the "hot debrief" after a mission or flight, a quick, immediate discussion of not just problems but also things that went well. By normalising the discussion of success factors, teams can reinforce those behaviours.

Learning from success often requires deliberate effort. After landing a first major client as a freelancer and completing the project smoothly, the instinct was to celebrate and move on. But pausing to ask *why* it went well revealed something crucial: proactive communication of expectations had set the tone for clarity and trust. That insight became a repeatable practice in future engagements. Celebration builds morale; analysis builds momentum. The same principle applies beyond work. In relationships, when a difficult conversation ends in understanding and closeness, it's worth asking what made it successful. Was it active listening? Empathy? Timing? Identifying those factors turns fleeting wins into lasting habits. Growth is forged in failure, and it is cemented when success is studied with the same rigour.

In summary, resilience learning is about learning from everything. Disruptions and failures scream for attention and indeed provide critical lessons on our vulnerabilities. Successes whisper more subtle lessons about our strengths and adaptive capacity. A resilient mindset listens to both. It treats every outcome as feedback. By learning from disruptions, we patch the holes in our hull; by learning from successes, we catch the wind in our sails. Both are necessary to navigate the uncertain waters ahead.

INDIVIDUAL, ORGANISATIONAL, AND SOCIETAL DIMENSIONS OF LEARNING

Learning to be resilient operates at multiple levels, from the deeply personal, to the teams and organisations we are part of, and all the way to communities and societies. In this section, we explore how learning processes play out in these different domains, and how they interconnect. Often, a lesson learned at the personal level can scale up, and lessons at the organisational or societal level can trickle down to individuals. Building a culture of resilience is thus a multi-layered effort.

Personal Learning and Growth

At the individual level, resilience through learning is deeply personal. It requires self-awareness, habits of reflection, and a deliberate effort to seek meaning. PTG is one powerful outcome of this process, where adversity becomes a catalyst for positive change. Research shows that individuals who experience PTG often report a heightened sense of personal strength (*If I survived that, I can handle other challenges*), deeper relationships and empathy, a reordering of life priorities, and a richer appreciation for everyday experiences. These shifts rarely happen automatically; they emerge through intentional introspection after the crisis.

One powerful strategy is journalling, a mirror for the mind. Writing about an experience, especially a painful one, slows down racing thoughts and begins to impose order on chaos. Patterns emerge on the page; causes and effects become clearer. This is why many therapists recommend expressive writing: it is a form of self-guided reflection that can unlock insight and emotional release. During a period of vulnerability, pages filled quickly, at first, raw venting, a torrent of "Why did this happen to me?" But over time, the tone shifted towards "What can I do with what happened to me?" That subtle change marked the turning point from rumination to learning. Reflection transformed adversity into a source of growth, proving that resilience is not just endurance, it is interpretation.

Another powerful practice is mindful self-talk or meditation. Mindfulness creates space between emotion and reaction, allowing thoughts and feelings to be observed without judgement. That space is where learning begins. After particularly stressful days, even a brief meditation can reveal what truly lies beneath the surface. Instead of the blunt "I'm angry", clarity emerges: "I'm angry because I felt excluded in that meeting, which touches an old insecurity". That insight points to action, perhaps voicing ideas more confidently, or addressing the insecurity itself. In this way, mindfulness turns raw emotion into usable information. And information is gold for resilience, because it prevents feelings from blindly dictating behaviour. Reflection transforms reaction into response, and response into growth.

Seeking feedback is one of the most powerful tools for personal growth. Blind spots often obscure how behaviour or performance affects others, and proactive feedback from trusted sources, such as a mentor, a friend, or a coach, can reveal lessons that self-reflection alone might miss. Early in a career, feedback after a chaotic project offered a hard truth: stress was spilling over to the team, creating a feedback loop of anxiety. It stung, but it was true. That insight reshaped leadership habits, reinforcing that resilience is about enduring pressure and managing signals under pressure to keep morale steady. A resilient individual is not someone who has all the answers internally, but someone willing to learn and course-correct. Often, that means listening, because resilience is strengthened not in isolation, but through openness to perspectives that challenge and refine our own.

The personal dimension of resilience often involves a shift in perspective that borders on the philosophical. Adversity forces a confrontation with fundamental questions: *What truly matters? What meaning can be drawn from this suffering?* For some, answers emerge through faith or spiritual traditions; for others, through frameworks like Stoicism or existentialism, which offer ways to interpret hardship. Whatever the source, finding a narrative that makes sense of what happened can transform pain into purpose.

In one case, the loss of a corporate role was reframed, not as a negation of worth, but as a necessary pivot towards a life aligned with core values. That shift turned a setback into a fork in the road, leading to greater meaning. Such reframing is more than optimism; it is a form of learning at the deepest level, about purpose, limits, and capacity for growth. When adversity becomes a lens for clarity rather than chaos, resilience moves beyond survival and becomes a foundation for transformation.

Organisational Learning and Adaptation

Organisations face their own challenges in embedding learning into their resilience fabric. A resilient organisation is often described as a *learning organisation*, one that continuously transforms itself through knowledge and adaptation. In practical terms, this means establishing processes, roles, and a culture that prioritises learning from every operation, project, and incident.

One key element is having formal mechanisms for learning, like the AARs and continuous improvement programmes we discussed. Companies that excel here treat every project completion or every incident resolution as the *beginning* of a learning process, not just an end. For example, some software companies practice *blameless post-mortems* after outages: when a service goes down and is restored, a meeting is held to dissect the incident without blaming individuals. The focus is on identifying system issues and improving procedures. The result is often a set of action items, maybe better monitoring tools to catch errors earlier, or cross-training team members to fill in key roles, which then get implemented. Over time, these organisations experience fewer severe outages because they don't make the same mistake twice (or at least strive not to). As one business resilience advisor succinctly put it: *Create a continuous improvement loop from each incident.* Every drill, disruption, or near miss should lead to adjustments that strengthen the organisation. In other words, lessons without action are not really learned; they must translate into changed behaviour or systems.

Organisational learning for resilience also heavily relies on culture. This cannot be overstated. If employees fear punishment for admitting mistakes, lessons will remain hidden. If teams are too siloed to communicate, lessons will stay localised and not benefit the wider organisation. Therefore, building a *resilient culture* often means fostering psychological safety (so people can speak up), encouraging cross-functional collaboration (so knowledge is shared), and supporting leadership that models learning behaviour (leaders who admit their own mistakes and demonstrate learning set a powerful example). A notable example is the aviation industry's safety culture: decades ago, cockpits were hierarchical and errors were often swept under the rug. After recognising that many accidents were preventable with better communication and learning, the industry embraced Crew Resource Management training, teaching junior crew to voice concerns and senior crew to listen. Incident reporting systems became non-punitive and encouraged reports of even minor issues. The result has been a dramatic improvement in safety. Airlines now share safety data globally, so a problem seen by one crew can become a lesson for all. This is organisational learning scaled up to an entire industry.

Another aspect of organisational learning is training and knowledge management. Organisations need to retain lessons learned even as personnel change. This is why standard operating procedures, training programmes, and knowledge repositories are important. However, they must be kept up-to-date. A manual written ten years ago about crisis response might be obsolete if none of the lessons from the last five crises made it into the pages. Some organisations address this by scheduling regular reviews of policies and including frontline employees in that review (since they know what really works and doesn't). Others use technology, for instance, maintaining a "knowledge wiki" that is continually edited with new insights. The goal is to avoid the trap of "corporate amnesia", where the organisation fails to remember past lessons and repeats mistakes a few years down the line.

Resilience at the organisational level includes more than recovering from setbacks; it also means codifying what works under pressure. Just as individuals benefit from understanding their strengths, organisations should identify and replicate success factors. After a major flood, one manufacturing site evacuated and restarted operations in record time with minimal loss. Investigation revealed the reason: an engaged emergency coordinator who ran monthly drills and kept response plans clear and current. That insight became an internal best practice, prompting other sites to adopt similar drill frequency and communication protocols.

This approach mirrors the practice in HROs of sharing "successful models of care". When one unit demonstrates exceptional performance, its methods are studied and disseminated widely. In essence, resilience is about fixing what failed, and it's about amplifying what worked, turning isolated wins into systemic strength.

Organisational resilience is built on institutionalising learning. Mechanisms like AARs, knowledge databases, and training ensure no lesson is lost. Cultural norms of openness and continuous improvement ensure that employees at all levels are engaged in the learning process. And leadership commitment ensures that learning translates into action and resources (e.g., investing in better tools if a lesson calls for it). An organisation that learns is one that doesn't just survive shocks; it adapts and often comes out stronger, with improved processes or new innovations sparked by overcoming difficulties. In a fast-changing world, an adaptive learning organisation can pivot where a rigid one might break.

Societal and Systemic Learning

When we zoom out to communities, cities, and societies, the stakes of learning from adversity become enormous. History provides plenty of examples of societies that learned and those that did not, often with life-and-death consequences. A classic case of societal learning is how cities adapt (or fail to adapt) after natural disasters. The maxim "Build Back Better" captures the ideal outcome: after a disaster, rebuild in a way that makes the community more resilient to future hazards. This implies that lessons from the disaster – perhaps building codes were inadequate, or social support networks were lacking – are identified and addressed in the rebuilding process.

Consider the example of earthquake-prone regions. In 1971, the San Fernando earthquake in California caused hospital buildings to collapse, leading to many deaths. It was a wake-up call that building standards were not sufficient for critical facilities. In response, California implemented much stricter hospital construction codes (the Alquist Hospital Seismic Safety Act). Similarly, after Mexico City's devastating 1985 earthquake, building codes were overhauled and enforcement strengthened. These changes are instances of societies learning, sometimes the hard way, and taking steps to prevent repeat tragedies. We see a similar pattern after catastrophic hurricanes or floods: levees are rebuilt taller, emergency alert systems improved, and land-use policies revised. The cycle is painfully clear: disaster, lessons identified, reforms (hopefully) improved resilience. However, the cycle doesn't always complete. Sometimes lessons are identified but not truly learned, as no action follows. A critical report might sit on a shelf due to political or financial hurdles. True societal learning requires the will to implement changes and the public support to sustain them.

Public health offers a recent and poignant example. The COVID-19 pandemic taught the global society harsh lessons about pandemic preparedness, supply chain dependencies (like for medical equipment), and the importance of clear communication. In its aftermath, many countries commissioned reviews. For instance, some countries established new pandemic response agencies or stockpiled critical supplies as a direct lesson learned from the shortages experienced in 2020. Whether these lessons will stick in the long run is an open question; collective memory can be short. Yet, resilience as evolution means each shock should leave lasting improvements. Some experts have stressed that as disasters become more frequent (due to climate change, etc.), we must consciously design mechanisms to ensure we "remember" and act on past lessons at the societal level. That could mean better knowledge-sharing between regions (so one city's experience benefits all), institutionalised disaster drills and education for citizens, and legislation that mandates resilience measures. For example, after numerous wildfires, some communities are adopting stricter land management and emergency evacuation policies. Those policies are essentially codified lessons, the hard-earned wisdom of what didn't work before and what might work better next time.

Societal learning also encompasses cultural attitudes. Sociologists have observed that some communities have a cultural narrative of resilience, "we have survived and will again", which can mobilise collective action and mutual aid after a crisis. This ties back to learning in the sense of shared story-making. A community that has come through adversity might memorialise it (through commemorations, stories, even art) in a way that educates future generations about both the tragedy and the recovery. This cultural memory can be a powerful teacher. It can remind people of risks ("Our grandparents remember when this river last flooded badly") and also of strengths ("Remember how the community came together after that flood"). Thus, storytelling and education become vehicles for societal resilience learning.

On a systems level (like national infrastructure or global networks), learning often requires coordination and overcoming complexity. Take the example of global aviation safety. A plane crash in one part of the world triggers alerts, and often aircraft manufacturers and airlines globally will learn from the investigation and modify things if needed. This is a highly interconnected form of learning, and one reason air travel has become so safe, because the whole system learns from each incident, not just the airline involved. Similarly, in the cybersecurity realm, when a major new threat or hack technique is discovered, that knowledge is disseminated among organisations to strengthen firewalls and protocols everywhere. We could say that in these cases, the system demonstrates resilience by learning: one node suffers an attack, but all nodes take heed and adapt.

It's worth noting that societal learning is not automatic. It often faces inertia and even resistance. People have different interpretations of events, politics can interfere, and resources might be scarce. But there are inspiring examples of learning leading to transformation. One such concept is "bouncing forward" instead of bouncing back. Thought leaders have suggested that communities should aim not merely to restore what was, but to create something better and more sustainable after a shock. For instance, after a financial crisis, some economies implemented regulatory reforms that made the financial system safer than before. After environmental crises, some societies shifted policies towards greener and more resilient practices (like moving a flood-prone town to higher ground rather than rebuilding in place). Bouncing forward implies a kind of collective wisdom: having survived the test, we now choose a wiser path. It is essentially resilience plus foresight, using the experience to leap ahead. One expert on disasters put it aptly: *We used to talk about coming back to what we were before. But in a time of climate change, we need to be talking about coming back to something that's more resilient than what we had before.* This captures the essence of learning at the societal level: it is about transformation and evolution for the collective good, informed by the scars and lessons of the past.

FROM RECOVERY TO TRANSFORMATION: RESILIENCE AS EVOLUTION

Recovery from adversity is a significant achievement on its own. Getting back on your feet after being knocked down, whether literally rebuilding a house after a hurricane or emotionally recovering from a personal loss, is an outcome to be proud of. But the resilience cycle doesn't end at recovery. The true promise of resilience lies in what comes *after* recovery: the potential for transformation. This is where learning plays the pivotal role. Learning is the alchemy which can turn the mere survival of a crisis into *growth,* because of that crisis.

Think of a simple timeline: *Pre-disruption → disruption → recovery →* (stop here, and you're essentially where you started). But add learning and insight, and the timeline becomes *Pre-disruption → disruption → recovery → improved state*. That improved state might be tangible (stronger infrastructure, new skills, better finances) or intangible (greater wisdom, confidence, closer relationships). Either way, it represents a form of evolution. The system, whether a person, or an organisation, or a city, has adapted and changed its baseline. It is now functioning on a different level, ideally a higher one, than before the adversity.

Throughout this chapter, we explored how reflective learning, feedback loops, and the integration of lessons drive resilience beyond mere recovery.

To bring it full circle, consider the opening story through this lens. A job loss stripped away stability and certainty. Recovery followed: new work, restored finances, a return to normalcy. If the story ended there, it would illustrate resilience as "bouncing back". But something deeper occurred: the disruption reshaped outlook and career trajectory. Reflection turned pain into insight, leading to a pivot towards work aligned with core values and a redefined approach to balance and relationships. This was not just recovery; it was reinvention. The experience planted a seed of growth that, when nurtured by learning, yielded something stronger and more purposeful. Resilience, at its highest form, is not about returning to what was. It is about moving forward with wisdom earned through struggle, bouncing forward, not just back.

In broader terms, resilience as evolution suggests that each cycle of challenge and recovery incrementally advances our capacity. A striking example outside the personal realm is technology and engineering. Think about how each failure in rocket development (and there have been many, famously with SpaceX and other private ventures) has led to innovations that made the next rockets better. Engineers treat failures as data, and with each iteration, the design evolves. After enough cycles, you have reusable rockets safely landing back on Earth, a feat that was science fiction a decade ago. The industry literally learned its way to transformation.

In community resilience, transformation might look like a town that was devastated by an earthquake rebuilding itself as a model of quake-proof urban design, or a community hit by economic collapse reinventing its local economy in a more sustainable, diversified way that thrives. These are not overnight changes; they are the result of collective learning and the will to implement bold changes coming out of hardship. In some cases, the transformation might be personal and collective at the same time; consider movements that arise from adversity, like Mothers Against Drunk Driving (MADD), which was started by a mother who lost her child to a drunk driver. Her personal tragedy led to a national movement that transformed laws and attitudes about drunk driving. Here, the recovery was not just private healing, but channelled into societal change. Such examples highlight how powerful learning can be when coupled with action: it can turn pain into progress.

A crucial insight for embracing resilience as evolution is maintaining a forward-looking mindset even as we recover. It is easy after a crisis to yearn for everything to go "back to normal". But if we simply restore the status quo without questioning it, we may be reinstating the same conditions that made us vulnerable in the first place. The transitional period of recovery is exactly when we should ask: *What do we want to be different now?* It is the time to implement those lessons we have catalogued. In a way, this is the final test of whether we truly learned. Learning is not complete until it's applied. A community might have stacks of "lessons learned" reports, but if no changes are made, then learning remains theoretical. Transformation requires that we apply the insight, update the building code, start the new training programme, invest in the stronger system, and practice the healthier habit. This is where courage is sometimes needed, because change can be uncomfortable. Yet, resilience inherently is about navigating discomfort and uncertainty. Applying lessons might mean taking some calculated risks or breaking from old habits, but those very moves can lead to lasting resilience dividends.

As we conclude this chapter and the resilience cycle, the tone is one of hope grounded in realism. Every disruption, by definition, causes loss or difficulty. We do not celebrate adversity itself. Rather, we acknowledge that within each adversity lies the possibility, not the guarantee, but the possibility, of growth. Whether that growth materialises depends on what we do in the aftermath. Reflective learning is the catalyst. It is how we distil meaning from chaos. It is how we remember what happened accurately and compassionately. It is how we honour those who suffered by doing better next time.

Resilience, then, is not a static shield against harm; it is a dynamic journey of becoming. It is iterative. We are always, in some sense, in training for the next challenge, even as we live through the current one. And with each iteration, if we embrace learning, we become not just tougher or "bounce-proof", but often wiser and more adaptive. A resilient system is a learning system. A resilient person is a learning person. The resilience cycle – prepare, absorb, recover, adapt, learn – repeats, but it does not spin in place. It spirals upward, each loop potentially elevating us to new capabilities or understanding.

Stepping off the ridge trail as the sun dips low, reflection settles on the spiral of the past year. The cycle turned, and the person who began it is not the same as the one who walks now. None of us are, after life's trials. The real question is whether those changes are integrated consciously and used constructively. As the path ahead unfolds, may we carry our journals of lessons, both literal and metaphorical, and use them as maps. The terrain will remain uncertain, but we are not static travellers. We are mapmakers and pathfinders, charting routes shaped by experience and reflection. In that sense, resilience is not merely about bouncing back; it is about journeying through and beyond, learning continuously, and becoming more with every step.

In the next part of this book, we will turn our attention to how all these pieces – anticipation, absorption, adaptation, and learning – come together in practice and how we can cultivate an enduring resilience mindset. But as we close this chapter on learning, let's remember: the story of resilience is still being written, every day, in our choices to learn and grow. The final stage of one resilience cycle plants the seeds of the next. Through learning, we transform recovery into evolution, and we ensure that every storm we weather leaves us not defeated but defined, and refined, by the wisdom it provided.

Table 5.1 presents an overview of the key themes and strategies discussed throughout this chapter, illustrating how resilience and learning operate at both individual and organisational levels. It consolidates practices such as

TABLE 5.1 Bouncing forward: how learning turns resilience into growth

THEMES	*KEY IDEAS*	*EXAMPLES/INSIGHTS*
Reflective Learning	Turning experiences into lessons through structured reflection	Journalling shifted from "Why did this happen?" to "What can I do with this?"
Feedback Loops	Continuous improvement by analysing wins and setbacks	Friday ritual: "What went well?" and "What can be better?" → improved communication
Learning from Success	Celebrate wins but analyse why they happened	Proactive expectation-setting led to smooth client project
Mindful Self-Talk	Observing emotions without judgement to uncover root causes	"I'm angry because I felt excluded" → actionable insight
Seeking Feedback	External perspectives reveal blind spots	Mentor noted stress signals → improved leadership resilience
Philosophical Reframing	Finding meaning in adversity to guide future choices	Job loss reframed as a pivot towards values-aligned work
Organisational Learning	Codifying and sharing success factors across teams	Flood response drills → best practice replicated across sites
Post-Traumatic Growth	Positive change through struggle and introspection	Greater strength, empathy, and life appreciation after crisis
Core Principle	Resilience = not just bouncing back, but bouncing forward	Growth through reflection, feedback, and meaning-making

reflective journalling, feedback loops, mindful self-talk, and philosophical reframing, alongside organisational approaches like codifying success factors. Each theme is paired with practical examples and insights, highlighting how these methods contribute to transforming adversity into growth and fostering a culture of continuous improvement.

Figure 5.1 illustrates resilience as a continuous cycle with five stages: Prepare, Absorb, Recover, Adapt, and Learn, where the learning phase closes the loop and feeds insights back into future preparedness.

Figure 5.2 illustrates the continuous improvement cycle that underpins resilience. It shows four stages: Action, Reflection, Adjustment, and New Action, connected in a loop. Each stage feeds into the next, creating a feedback mechanism that transforms experiences into adaptive strategies for future challenges.

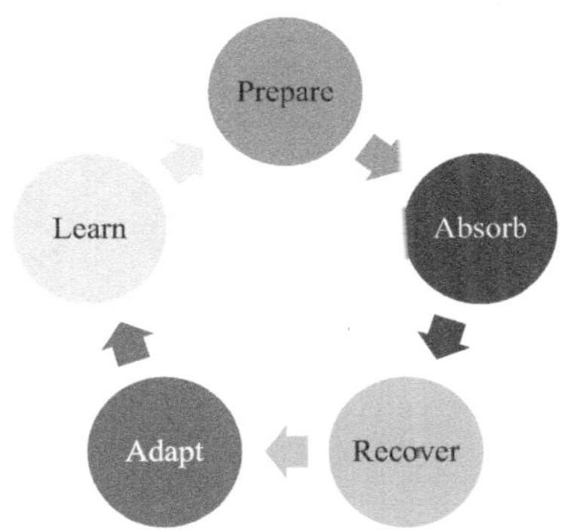

FIGURE 5.1 The resilience learning cycle.

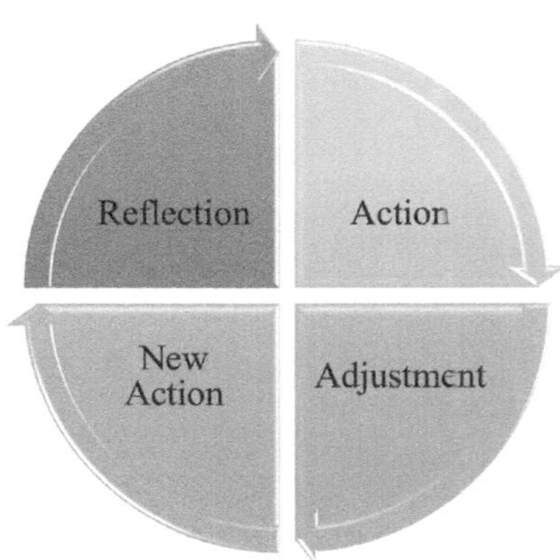

FIGURE 5.2 Feedback loops driving resilience.

REFERENCES

1. Housel, M., 2020. *The psychology of money: Timeless lessons on wealth, greed, and happiness*. Harriman House Limited.
2. Rogers, P., 2020. The evolution of resilience. *Connections*, 19(3), pp.13–32.
3. de Oliveira Teixeira, E. and Werther Jr, W.B., 2013. Resilience: Continuous renewal of competitive advantages. *Business Horizons*, 56(3), pp.333–342.
4. Sterman, B., 2012. *Rarest blue: The remarkable story of an ancient color lost to history and rediscovered*. Rowman & Littlefield.
5. Tedeschi, R.G. and Calhoun, L.G., 2004. A clinical approach to posttraumatic growth. *Positive Psychology in Practice*, pp.405–419.
6. Calhoun, L.G. and Tedeschi, R.G., 2001. Posttraumatic growth: The positive lessons of loss. In *Meaning reconstruction & the experience of loss*. Washington, D.C.: American Psychological Association, pp.157–172.

Index